Lecture Notes in Statistics 118

Edited by P. Bickel, P. Diggle, S. Fienberg, K. Krickeberg,
I. Olkin, N. Wermuth, S. Zeger

Springer
*New York
Berlin
Heidelberg
Barcelona
Hong Kong
London
Milan
Paris
Singapore
Tokyo*

Radford M. Neal

Bayesian Learning for Neural Networks

 Springer

Radford M. Neal
Department of Statistics and
Department of Computer Science
University of Toronto
Toronto, Ontario
Canada M5S 1A4

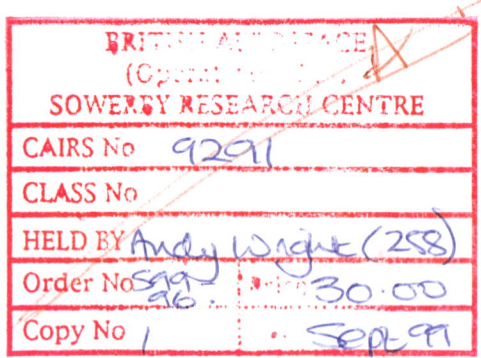

CIP data available.
Printed on acid-free paper.

© 1996 Springer-Verlag New York, Inc.
All rights reserved. This work may not be translated or copied in whole or in part without the written permission of the publisher (Springer-Verlag New York, Inc., 175 Fifth Avenue, New York, NY 10010, USA), except for brief excerpts in connection with reviews or scholarly analysis. Use in connection with any form of information storage and retrieval, electronic adaptation, computer software, or by similar or dissimilar methodology now known or hereafter developed is forbidden.
The use of general descriptive names, trade names, trademarks, etc., in this publication, even if the former are not especially identified, is not to be taken as a sign that such names, as understood by the Trade Marks and Merchandise Marks Act, may accordingly be used freely by anyone.

Camera ready copy provided by the author.
Printed and bound by Electronic Printing Inc., Plainview, NY.
Printed in the United States of America.

9 8 7 6 5 4 3

ISBN 0-387-98222-1 Springer-Verlag New York Berlin Heidelberg SPIN 10727303

Preface

This book, a revision of my Ph.D. thesis,[1] explores the Bayesian approach to learning flexible statistical models based on what are known as "neural networks". These models are now commonly used for many applications, but understanding why they (sometimes) work well, and how they can best be employed is still a matter for research. My aim in the work reported here is two-fold — to show that a Bayesian approach to learning these models can yield theoretical insights, and to show also that it can be useful in practice. The strategy for dealing with complexity that I advocate here for neural network models can also be applied to other complex Bayesian models, as can the computational methods that I employ.

In Chapter 1, I introduce the Bayesian framework for learning, the neural network models that will be examined, and the Markov chain Monte Carlo methods on which the implementation is based. This presentation presupposes only that the reader possesses a basic statistical background.

Chapter 1 also introduces the major themes of this book, which involve two fundamental characteristics of Bayesian learning. First, Bayesian learning starts with a prior probability distribution for model parameters, which is supposed to capture our beliefs about the problem derived from background knowledge. Second, Bayesian predictions are not based on a single estimate for the model parameters, but rather are found by integrating the

---

[1] *Bayesian Learning for Neural Networks*, Department of Computer Science, University of Toronto, 1995.

model's predictions with respect to the posterior parameter distribution that we obtain when we update the prior to take account of the data. For neural network models, both these aspects present difficulties — the prior over network parameters has no obvious relation to any prior knowledge we are likely to have, and integration over the posterior distribution is computationally very demanding.

I address the first of these problems in Chapter 2, by defining classes of prior distributions for network parameters that reach sensible limits as the size of the network goes to infinity. In this limit, the properties of these priors can be elucidated. Some priors converge to Gaussian processes, in which functions computed by the network may be smooth, Brownian, or fractionally Brownian. Other priors converge to non-Gaussian stable processes. Interesting effects are obtained by combining priors of both sorts in networks with more than one hidden layer. This work shows that within the Bayesian framework there is no theoretical need to limit the complexity of neural network models. Indeed, limiting complexity is likely to conflict with our prior beliefs, and can therefore be justified only to the extent that it is necessary for computational reasons.

The computational problem of integrating over the posterior distribution is addressed in Chapter 3, using Markov chain Monte Carlo methods. I demonstrate that the hybrid Monte Carlo algorithm, originally developed for applications in quantum chromodynamics, is superior to the methods based on simple random walks that are widely used in statistical applications at present. The hybrid Monte Carlo method makes the use of complex Bayesian network models possible in practice, though the computation time required can still be substantial.

In Chapter 4, I use a hybrid Monte Carlo implementation to test the performance of Bayesian neural network models on several synthetic and real data sets. Good results are obtained on small data sets when large networks are used in conjunction with priors designed to reach limits as network size increases, confirming that with Bayesian learning one need not restrict the complexity of the network based on the size of the data set. A Bayesian approach is also found to be effective in automatically determining the relevance of inputs.

Finally, in Chapter 5, I draw some conclusions from this work, and briefly discuss related work by myself and others since the completion of the original thesis.

Readers interested in pursuing research in this area may obtain free software implementing the methods, as described in Appendix B. One should note, however, that this software is not intended for use in routine data analysis. The software is also designed only for use on Unix systems.

Of the many people who have contributed to this work, I would like first of all to thank my thesis advisor, Geoffrey Hinton. His enthusiasm for understanding learning, his openness to new ideas, and his ability to provide insightful criticism have made working with him a joy. I am also fortunate to have been part of the research group he has led, and of the wider AI group at the University of Toronto. I would particularly like to thank fellow students Richard Mann, Carl Rasmussen, and Chris Williams for their helpful comments on this work and its precursors. My thanks also go to the present and former members of my Ph.D. committee, Mike Evans, Scott Graham, Rudy Mathon, Demetri Terzopoulos, and Rob Tibshirani.

I am especially pleased to thank David MacKay, whose work on Bayesian learning and its application to neural network models has been an inspiration to me. He has also contributed much to this work through many conversations and e-mail exchanges, which have ranged from the philosophy of Bayesian inference to detailed comments on presentation. I have benefited from discussions with other researchers as well, in particular, Wray Buntine, Brian Ripley, Hans Henrik Thodberg, and David Wolpert.

This work was funded by the Natural Sciences and Engineering Research Council of Canada and by the Information Technology Research Centre. For part of my studies, I was supported by an Ontario Government Scholarship.

Contents

Preface   iii

1 Introduction   1
    1.1 Bayesian and frequentist views of learning 3
        1.1.1 Models and likelihood 3
        1.1.2 Bayesian learning and prediction 4
        1.1.3 Hierarchical models 6
        1.1.4 Learning complex models 7
    1.2 Bayesian neural networks 10
        1.2.1 Multilayer perceptron networks 10
        1.2.2 Selecting a network model and prior 14
        1.2.3 Automatic Relevance Determination (ARD) models . 15
        1.2.4 An illustration of Bayesian learning for a neural net . 17
        1.2.5 Implementations based on Gaussian approximations . 19
    1.3 Markov chain Monte Carlo methods 22
        1.3.1 Monte Carlo integration using Markov chains 23
        1.3.2 Gibbs sampling 25
        1.3.3 The Metropolis algorithm 26
    1.4 Outline of the remainder of the book 28

2 Priors for Infinite Networks — 29

- 2.1 Priors converging to Gaussian processes 31
  - 2.1.1 Limits for Gaussian and other priors with finite variance — 32
  - 2.1.2 Priors that lead to smooth and Brownian functions . — 34
  - 2.1.3 Covariance functions of Gaussian priors . . . . — 37
  - 2.1.4 Fractional Brownian priors . . . . — 39
  - 2.1.5 Networks with more than one input . . . . — 40
- 2.2 Priors converging to non-Gaussian stable processes . . . . . . — 43
  - 2.2.1 Limits for priors with infinite variance . . . . — 43
  - 2.2.2 Properties of non-Gaussian stable priors . . . . — 45
- 2.3 Priors for nets with more than one hidden layer . . . . — 48
- 2.4 Hierarchical models . . . . — 51

3 Monte Carlo Implementation — 55

- 3.1 The hybrid Monte Carlo algorithm . . . . — 56
  - 3.1.1 Formulating the problem in terms of energy . . . . . . — 57
  - 3.1.2 The stochastic dynamics method . . . . — 58
  - 3.1.3 Hybrid Monte Carlo . . . . — 60
- 3.2 An implementation of Bayesian neural network learning . . . — 63
  - 3.2.1 Gibbs sampling for hyperparameters . . . . — 66
  - 3.2.2 Hybrid Monte Carlo for network parameters . . . . . — 68
  - 3.2.3 Verifying correctness . . . . — 73
- 3.3 A demonstration of the hybrid Monte Carlo implementation — 74
  - 3.3.1 The robot arm problem . . . . — 75
  - 3.3.2 Sampling using the hybrid Monte Carlo method . . . — 76
  - 3.3.3 Making predictions . . . . — 84
  - 3.3.4 Computation time required . . . . — 87
- 3.4 Comparison of hybrid Monte Carlo with other methods . . . — 88
- 3.5 Variants of hybrid Monte Carlo . . . . — 91
  - 3.5.1 Computation of trajectories using partial gradients . — 91
  - 3.5.2 The windowed hybrid Monte Carlo algorithm . . . . . — 95
  - 3.5.3 Hybrid Monte Carlo with persistent momentum . . . — 97

4 Evaluation of Neural Network Models — 99

- 4.1 Network architectures, priors, and training procedures . . . . — 100
- 4.2 Tests of the behaviour of large networks . . . . — 102
  - 4.2.1 Theoretical expectations concerning large networks . . — 103

		4.2.2 Tests of large networks on the robot arm problem	104
	4.3	Tests of Automatic Relevance Determination	113
		4.3.1 Procedures for evaluating ARD models	114
		4.3.2 Tests of ARD on the noisy LED display problem	116
		4.3.3 Tests of ARD on the robot arm problem	122
	4.4	Tests of Bayesian models on real data sets	126
		4.4.1 Methodology for comparing learning procedures	126
		4.4.2 Tests on the Boston housing data	127
		4.4.3 Tests on the forensic glass data	136
5	**Conclusions and Further Work**		**145**
	5.1	Priors for complex models	145
	5.2	Hierarchical Models — ARD and beyond	147
	5.3	Implementation using hybrid Monte Carlo	150
	5.4	Evaluating performance on realistic problems	152
A	**Details of the Implementation**		**153**
	A.1	Specifications	153
		A.1.1 Network architecture	153
		A.1.2 Data models	155
		A.1.3 Prior distributions for parameters and hyperparameters	156
		A.1.4 Scaling of priors	159
	A.2	Conditional distributions for hyperparameters	159
		A.2.1 Lowest-level conditional distributions	160
		A.2.2 Higher-level conditional distributions	160
	A.3	Calculation of derivatives	161
		A.3.1 Derivatives of the log prior density	162
		A.3.2 Log likelihood derivatives with respect to unit values	162
		A.3.3 Log likelihood derivatives with respect to parameters	163
	A.4	Heuristic choice of stepsizes	164
	A.5	Rejection sampling from the prior	166
B	**Obtaining the software**		**169**
	Bibliography		**171**
	Index		**177**

List of Figures

1.1	A multilayer perceptron network	11
1.2	An illustration of Bayesian inference for a neural network	18
2.1	Convergence of network priors to a Gaussian process	33
2.2	Functions drawn from Gaussian priors for networks of step-function units	35
2.3	Functions drawn from Gaussian priors for networks of tanh hidden units	37
2.4	Functions drawn from fractional Brownian priors	41
2.5	Behaviour of $D(x-s/2, x+s/2)$ for different sorts of functions	41
2.6	Functions of two inputs drawn from Gaussian priors	42
2.7	Functions drawn from Cauchy priors	46
2.8	Functions of two inputs drawn from non-Gaussian priors	47
2.9	Functions computed from networks with several hidden layers	49
2.10	Functions drawn from a combined Gaussian and non-Gaussian prior	50
3.1	Sampling using the Langevin and hybrid Monte Carlo methods	63
3.2	Progress of hybrid Monte Carlo runs in the initial phase	78

xiv   List of Figures

3.3 Error in energy for trajectories computed with different step-sizes 80
3.4 Degree of correlation along a trajectory 80
3.5 Progress of hybrid Monte Carlo runs in the sampling phase .. 82
3.6 Autocorrelations for different trajectory lengths 83
3.7 Predictive distribution from Monte Carlo data 85
3.8 Average test error on the robot arm problem with different implementations 86
3.9 Progress of simple Metropolis and Langevin methods in the sampling phase 89
3.10 Error in energy for trajectories computed using partial gradients 94
3.11 Difference in free energy for windowed trajectories 94

4.1 Computational details for experiments on networks of varying size 107
4.2 Results on the robot arm problem with networks of varying size 109
4.3 Predictive distributions obtained using networks of varying size 110
4.4 Results of maximum likelihood learning with networks of varying size 112
4.5 Digit patterns for the noisy LED display problem 116
4.6 Results on the noisy LED display problem 120
4.7 Relevant and irrelevant input weight magnitudes for the LED display problem 121
4.8 Input weight magnitudes for the robot arm problem with and without ARD 124
4.9 Descriptions of inputs for the Boston housing problem 128
4.10 Results of preliminary tests on the Boston housing data ... 130
4.11 Cross-validation assessments on the Boston housing data ... 134
4.12 Networks and priors tested on the forensic glass data 139
4.13 Results on the forensic glass data 140
4.14 Effect of vague priors in the forensic glass problem 142

5.1 A hierarchical network model capable of finding additive structure 149

Chapter 1

Introduction

This book develops the Bayesian approach to learning for neural networks by examining the meaning of the prior distributions that are the starting point for Bayesian learning, by showing how the computations required by the Bayesian approach can be performed using Markov chain Monte Carlo methods, and by evaluating the effectiveness of Bayesian methods on several real and synthetic data sets. This work has practical significance for modeling data with neural networks. From a broader perspective, it shows how the Bayesian approach can be successfully applied to complex models, and in particular, challenges the common notion that one must limit the complexity of the model used when the amount of training data is small. I begin here by introducing the Bayesian framework, discussing past work on applying it to neural networks, and reviewing the basic concepts of Markov chain Monte Carlo implementation.

Our ability to learn from observation is our primary source of knowledge about the world. We learn to classify objects — to tell cats from dogs, or an 'A' from a 'B' — on the basis of instances presented to us, not by being given a set of classification rules. Experience also teaches us how to predict events — such as a rainstorm, or a family quarrel — and to estimate unseen quantities — such as when we judge the likely weight of an object from its size and appearance. Without this ability to learn from empirical data, we would be unable to function in daily life.

Theories and methodologies of learning are interesting from a number of perspectives. Psychologists try to model the learning abilities of humans and other animals, and to formulate high-level theories of how learning operates, while neurobiologists try to understand the biological mechanisms of learning at a lower level. Workers in artificial intelligence would like to understand in a more general way how learning is possible in a computational system, and engineers try to apply such insights to produce useful devices. Statisticians develop methods of inference from data that for certain tasks are more reliable and more sensitive than unaided common sense. Philosophers would like to understand the fundamental nature and justification of inductive learning.

The work I report in this book is aimed primarily at engineering applications. The "neural network" models used are designed for predicting an unknown category or quantity on the basis of known attributes. Such models have been applied to a wide variety of tasks, such as recognizing hand-written digits (Le Cun, *et al* 1990), determining the fat content of meat (Thodberg 1996), and predicting energy usage in buildings (MacKay 1993). Some of the methods I develop here may also have uses in statistical inference for scientific applications, where the objective is not only to predict well, but also to obtain insight into the nature of the process being modeled. Although neural networks were originally intended as abstract models of the brain, I do not investigate whether the models and algorithms I develop here might have a role in neural or psychological models.

The work I describe does have wider implications for the philosophy of induction, and its applications to artificial intelligence and statistics. The Bayesian framework for learning, on which this work is based, has been the subject of controversy for several hundred years. It is clear that the merits of Bayesian and competing approaches will not be settled by philosophical disputation, but only by demonstrations of effectiveness in practical contexts. I hope that the work I report here will contribute in this respect. In another direction, the infinite network models I discuss challenge common notions regarding the need to limit the complexity of models, and raise questions about the meaning and utility of "Occam's Razor" within the Bayesian framework.

The next section introduces the Bayesian view of learning in a general context. I then describe past work on applying the Bayesian framework to learning for neural networks, and indicate how this work will contribute to this approach in two respects — first, by examining the meaning for neural network models of the prior distribution that is the starting point for Bayesian learning, and second, by showing how the posterior distribution needed for making predictions in the Bayesian framework can be obtained using Markov chain Monte Carlo methods. To provide a foundation for the latter work, I also briefly review the basics of the Markov chain Monte Carlo method.

1.1 Bayesian and frequentist views of learning

The statistical methodology of *Bayesian learning* is distinguished by its use of probability to express all forms of uncertainty. Learning and other forms of inference can then be performed by what are in theory simple applications of the rules of probability. The results of Bayesian learning are expressed in terms of a probability distribution over all unknown quantities. In general, these probabilities can be interpreted only as expressions of our degree of belief in the various possibilities.

In contrast, the conventional *frequentist* approach to statistics uses probabilities only to represent the long-run frequencies of the outcomes of repeatable experiments. A frequentist strategy for learning takes the form of an *estimator* for unknown quantities, which one tries to show will usually produce good results.

To illustrate the difference between Bayesian and frequentist learning, consider tossing a coin of unknown properties. There is an irreducible uncertainty regarding the outcome of each toss, which can be expressed by saying that the coin has a certain probability of landing heads rather than tails. Since the properties of the coin are uncertain, however, we do not know what this probability of heads is (it might not be one-half). A Bayesian will express this uncertainty using a probability distribution over possible values for the unknown probability of the coin landing heads, and will update this distribution using the rules of probability theory as the outcome of each toss becomes known. To a frequentist, such a probability distribution makes no sense, since there is only one coin in question, and its properties are in fact fixed. The frequentist will instead choose some estimator for the unknown probability of heads, such as the frequency of heads in past tosses, and try to show that this estimator is good according to some criterion.

Introductions to Bayesian statistics are provided by Press (1989), Robert (1995), and Schmitt (1969); Berger (1985), Bernardo and Smith (1994), Box and Tiao (1973), DeGroot (1970), and Gelman, Carlin, Stern, and Rubin (1995) offer more advanced treatments. Barnett (1982) presents a comparative view of different approaches to statistical inference. Unfortunately, these books do not deal much with complex models of the sort that are the subject of this book.

1.1.1 Models and likelihood

Consider a series of quantities, $x^{(1)}$, $x^{(2)}$, ..., generated by some process in which each $x^{(i)}$ is independently subject to random variation. We can define a *probabilistic model* for this random process, in which a set of unknown *model parameters*, θ, determine the probability distributions of the $x^{(i)}$. Such probabilities, or probability densities, will be written in the form

$P(x^{(i)} \mid \theta)$. In the coin tossing example, the $x^{(i)}$ are the results of the tosses (heads or tails), and θ is the unknown probability of the coin landing heads; we then have $P(x^{(i)} \mid \theta) = [\theta$ if $x^{(i)} =$ heads; $1-\theta$ if $x^{(i)} =$ tails]. Another simple situation is when the $x^{(i)}$ are real-valued quantities assumed to have a Gaussian distribution, with mean and standard deviation given by $\theta = \{\mu, \sigma\}$. In this case, $P(x^{(i)} \mid \mu, \sigma) = \exp(-(x^{(i)} - \mu)^2 / 2\sigma^2) / \sqrt{2\pi}\sigma$.

Learning about θ is possible if we have observed the values of some of the $x^{(i)}$, say $x^{(1)}, \ldots, x^{(n)}$. For Bayesian as well as many frequentist approaches, the impact of these observations is captured by the *likelihood function*, $L(\theta) = L(\theta \mid x^{(1)}, \ldots, x^{(n)})$, which gives the probability of the observed data as a function of the unknown model parameters:

$$L(\theta) = L(\theta \mid x^{(1)}, \ldots, x^{(n)})$$
$$\propto P(x^{(1)}, \ldots, x^{(n)} \mid \theta) = \prod_{i=1}^{n} P(x^{(i)} \mid \theta) \quad (1.1)$$

This definition is written as a proportionality because all that matters is the relative values of $L(\theta)$ for different values of θ.

In the method of *maximum likelihood*, the unknown parameters are estimated by the value, $\widehat{\theta}$, that maximizes the likelihood, $L(\theta \mid x^{(1)}, \ldots, x^{(n)})$. In the coin tossing problem, the maximum likelihood estimate for θ turns out to be the frequency of heads among $x^{(1)}, \ldots, x^{(n)}$. For many models, use of the maximum likelihood estimate can be justified in frequentist terms on the basis that it has certain desirable properties, such as convergence to the true value as the amount of observational data increases. The maximum likelihood method does not always work well, however. When it doesn't, the method of *maximum penalized likelihood* estimation is sometimes better. This procedure estimates θ by the value that maximizes the product of the likelihood and a penalty function, which may be designed to "regularize" the estimate, perhaps by favouring values that are in some sense less "extreme".

In engineering applications, we are usually not interested in the value of θ itself, but rather in the value of some quantity that may be observed in the future, say $x^{(n+1)}$. In a frequentist context, the most obvious way of predicting such quantities is to use the estimated value for θ, basing our prediction on $P(x^{(n+1)} \mid \widehat{\theta})$. More sophisticated methods that take account of the remaining uncertainty in θ are also possible.

1.1.2 Bayesian learning and prediction

The result of Bayesian learning is a probability distribution over model parameters that expresses our beliefs regarding how likely the different parameter values are. To start the process of Bayesian learning, we must define a *prior distribution*, $P(\theta)$, for the parameters, that expresses our initial beliefs about their values, before any data has arrived. When we observe $x^{(1)}, \ldots, x^{(n)}$, we update this prior distribution to a *posterior distribution*, using Bayes' Rule:

$$P(\theta \mid x^{(1)}, \ldots, x^{(n)}) = \frac{P(x^{(1)}, \ldots, x^{(n)} \mid \theta) P(\theta)}{P(x^{(1)}, \ldots, x^{(n)})} \quad (1.2)$$

$$\propto L(\theta \mid x^{(1)}, \ldots, x^{(n)}) P(\theta) \quad (1.3)$$

The posterior distribution combines the likelihood function, which contains the information about θ derived from observation, with the prior, which contains the information about θ derived from our background knowledge. The introduction of a prior is a crucial step that allows us to go from a likelihood function to a probability distribution, and thereby allows learning to be performed using the apparatus of probability theory. The prior is also a common focus for criticism of the Bayesian approach, as some people view the choice of a prior as being arbitrary.

In the coin tossing example, we might start with a uniform prior for θ, the probability of heads. As we see the results of more and more tosses, the posterior distribution obtained by combining this prior with the likelihood function will become more and more concentrated in the vicinity of the value corresponding to the observed frequency of heads.

To predict the value of an unknown quantity, $x^{(n+1)}$, a Bayesian integrates the predictions of the model with respect to the posterior distribution of the parameters, giving

$$P(x^{(n+1)} \mid x^{(1)}, \ldots, x^{(n)}) = \int P(x^{(n+1)} \mid \theta) P(\theta \mid x^{(1)}, \ldots, x^{(n)}) d\theta \quad (1.4)$$

This *predictive distribution* for $x^{(n+1)}$ given $x^{(1)}, \ldots, x^{(n)}$ is the complete Bayesian inference regarding $x^{(n+1)}$, which can be used for many purposes, depending on the needs of the user. The ability to produce such a distribution is one advantage of the Bayesian approach.

In some circumstances we may need to make a single-valued guess at the value of $x^{(n+1)}$. The best way to do this depends on our *loss function*, $\ell(x, \widehat{x})$, which expresses our judgement of how bad it is to guess $\widehat{x}$ when the real value is x. For *squared error loss*, $\ell(x, \widehat{x}) = (x - \widehat{x})^2$, guessing the mean of the predictive distribution minimizes the expected loss. For *absolute error loss*, $\ell(x, \widehat{x}) = |x - \widehat{x}|$, the best strategy is to guess the median of the predictive distribution. For discrete-valued x, we might choose to use *0–1 loss*, which is zero if we guess correctly, and one if we guess incorrectly. The optimal strategy is then to guess the mode of the predictive distribution.

In the coin tossing example, if we use a uniform prior for the probability of heads, the Bayesian prediction for the result of toss $n+1$ given the results of the first n tosses turns out to be $P(x^{(n+1)} \mid x^{(1)}, \ldots, x^{(n)}) = [(h+1)/(n+2)$ if $x^{(n+1)} =$ heads; $(t+1)/(n+2)$ if $x^{(n+1)} =$ tails], where h and t are the numbers of heads and tails amongst $x^{(1)}, \ldots, x^{(n)}$. If we have a 0–1 loss function, we should guess that $x^{(n+1)}$ will be a head if $h > t$, but guess tails if $t > h$ (if $h = t$, both guesses are equally good). This is of course just what we would expect, and is also what we would be led to do using the maximum likelihood estimate of $\hat{\theta} = h/n$.

However, even in this simple problem we can see the effect of prediction by integration rather than maximization if we consider more complicated actions. We might, for example, have the option of not guessing at all, and may wish to make a guess only if we are nearly certain that we will be right. If we have tossed the coin twice, and each time it landed heads, naive application of maximum likelihood will lead us to conclude that the coin is certain to land heads on the next toss, since $\hat{\theta} = 1$. The Bayesian prediction with a uniform prior is a more reasonable probability of 3/4 for heads, which might not be high enough to prompt us to guess. The Bayesian procedure avoids jumping to conclusions by considering not just the value of θ that explains the data best, but also other values of θ that explain the data reasonably well, and hence also contribute to the integral of equation (1.4).

The formation of a predictive distribution by the integration of equation (1.4) is at the heart of Bayesian inference. Unfortunately, it is often the source of considerable computational difficulties as well. Finding the single value of θ with maximum posterior probability density is usually much easier. Use of this *maximum a posteriori probability (MAP)* estimate is sometimes described as a Bayesian method, but this characterization is inaccurate except when one can argue that the result of using this single value approximates the integral of equation (1.4). In general, this is not true — indeed, the MAP estimate can be shifted almost anywhere simply by switching to a new parameterization of the model that is equivalent to the old, but related to it by a nonlinear transformation. MAP estimation is better characterized as a form of maximum penalized likelihood estimation, with the penalty being the prior density of the parameter values in some preferred parameterization.

1.1.3 Hierarchical models

In the previous section, a common parameter, θ, was used to model the distribution of many observable quantities, $x^{(i)}$. In the same way, when the parameter has many components, $\theta = \{\theta_1, \ldots, \theta_p\}$, it may be useful to specify their joint prior distribution using a common *hyperparameter*, say γ, which is given its own prior. Schemes such as this are known as *hierarchical models*, and may be carried to any number of levels.

If the θ_k are independent given γ, we will have

$$P(\theta) = P(\theta_1,\ldots,\theta_p) = \int P(\gamma) \prod_{k=1}^{p} P(\theta_k \mid \gamma) \, d\gamma \qquad (1.5)$$

Mathematically, we could have dispensed with γ and simply written a direct prior for θ corresponding to the result of this integration. (In general the θ_k will not be independent in this direct prior.) The formulation using a hyperparameter may be much more intelligible, however. The situation is the same at the lower level — we *could* integrate over θ to produce a specification of the model in terms of a direct prior for the observable variables $x^{(1)}, x^{(2)},\ldots$, but most models lose their intuitive meaning when expressed in this form.

To give a simple example, suppose the observable variables are the weights of various dogs, each classified according to breed, and that θ_k is the mean weight for breed k, used to specify a Gaussian distribution for weights of dogs of that breed. Rather than using the same prior for each θ_k, independently, we could instead give each a Gaussian prior with a mean of γ, and then give γ itself a prior as well. The effect of this hierarchical structure can be seen by imagining that we have observed dogs of several breeds and found them all to be heavier than expected. Rather than stubbornly persisting with our underestimates for every new breed we encounter, we will instead adjust our idea of how heavy dogs are in general by changing our view of the likely value of the hyperparameter γ. We will then start to expect even dogs of breeds that we have never seen before to be heavier than we would have expected at the beginning.

One way of avoiding needless intellectual effort when defining a hierarchical model is to give the top-level hyperparameters prior distributions that are very vague, or even improper (i.e. have density functions whose integrals diverge). Often, the data is sufficiently informative that the posterior distributions of such hyperparameters become narrow despite the vagueness of the prior. Moreover, the posterior would often change very little even if we were to expend the effort needed to define a more specific prior for the hyperparameters that expressed our exact beliefs. One should not use vague or improper priors recklessly, however, as they are not always innocuous.

1.1.4 Learning complex models

"Occam's Razor" — the principle that we should prefer simple to complex models when the latter are not necessary to explain the data — is often held to be an essential component of inductive inference. In scientific contexts, its merits seem clear. In the messy contexts typical of engineering applications, its meaning and utility are less obvious. For example, we do not expect that there is any simple procedure for recognizing handwriting. The shapes of letters are arbitrary; they are written in many styles, whose characteristics are more a matter of fashion than of theory; stains and dirt may appear, and must somehow be recognized as not being part of the letters. Indeed, there is no reason to suppose that there is any limit to the complications involved in this task. It will always be possible to improve performance at least a bit by taking account of further rare writing styles, by modeling the shapes of the less common forms of ink blots, or by employing a deeper analysis of English prose style in order to make better guesses for smudged letters.

It is a common belief, however, that restricting the complexity of the models used for such tasks is a good thing, not just because of the obvious computational savings from using a simple model, but also because it is felt that too complex a model will *overfit* the training data, and perform poorly when applied to new cases. This belief is certainly justified if the model parameters are estimated by maximum likelihood. I will argue here that concern about overfitting is not a good reason to limit complexity in a Bayesian context.

One way of viewing the overfitting problem from a frequentist perspective is as a trade-off between the bias and the variance of an estimator, both of which contribute to the expected squared error when using the estimate to predict an observable quantity (Geman, Bienenstock, and Doursat 1992). These quantities may depend on the true underlying process, and reflect expectations with respect to the random generation of training data from this process. The bias of an estimator measures any systematic tendency for it to deliver the wrong answer; the variance measures the degree to which the estimate is sensitive to the randomness of the training examples.

One strategy for designing a learning procedure is to try to minimize the sum of the (squared) bias and the variance (note, however, that the procedure that minimizes this sum depends on the unknown true process). Since reducing bias often increases variance, and vice versa, minimizing their sum will generally require a trade-off. Controlling the complexity of the model is one way to perform this trade-off. A complex model that is flexible enough to represent the true process can have low bias, but may suffer from high variance, since its flexibility also lets it fit the random variation in the training data. A simple model will have high bias, unless the true process is really that simple, but will have lower variance. There are also other ways to trade off bias and variance, such as by use of a penalty function, but adjusting the model complexity is perhaps the most common method.

This strategy leads to a choice of model that varies with the amount of training data available — the more data, the more complex the model used. In this way, one can sometimes guarantee that the performance achieved will approach the optimum as the size of the training set goes to infinity, as the bias will go down with increasing model complexity, while the variance will also go down due to the increasing amounts of data (provided the accompanying increase in model complexity is sufficiently slow). Rules of thumb are sometimes used to decide how complex a model should be used with a given size training set (e.g. limit the number of parameters to some fraction of the number of data points). More formal approaches of this sort include the "method of sieves" (Grenander 1981) and "structural risk minimization" (Vapnik 1982).

From a Bayesian perspective, adjusting the complexity of the model based on the amount of training data makes no sense. A Bayesian defines a model, selects a prior, collects data, computes the posterior, and then makes predictions. There is no provision in the Bayesian framework for changing the model or the prior depending on how much data was collected. If the model and prior are correct for a thousand observations, they are correct for ten observations as well (though the impact of using an incorrect prior might be more serious with fewer observations). In practice, we might sometimes switch to a simpler model if it turns out that we have little data, and we feel that we will consequently derive little benefit from using a complex, computationally expensive model, but this would be a concession to practicality, rather than a theoretically desirable procedure.

For problems where we do not expect a simple solution, the proper Bayesian approach is therefore to use a model of a suitable type that is as complex as we can afford computationally, regardless of the size of the training set. Young (1977), for example, uses polynomial models of indefinitely high order. I have applied mixture models with infinite numbers of components to small data sets (Neal 1992a); the infinite model can in this case be implemented with finite resources. Nevertheless, this approach to complexity has not been widely appreciated — at times, not even in the Bayesian literature.

I hope that the work described in this book will help increase awareness of this view of complexity. In addition to the philosophical interest of the idea, avoiding restrictions on the complexity of the model should have practical benefits in allowing the maximum information to be extracted from the data, and in producing a full indication of the uncertainty in the predictions.

In light of this discussion, we might ask whether Occam's Razor is of any use to Bayesians. Perhaps. In some scientific applications, simple explanations may be quite plausible. Jeffreys and Berger (1992) give an example of this sort, illustrating that Bayesian inference embodies an automatic preference for such simple hypotheses. The same point is discussed by MacKay (1992a) in the context of more complex models, where "simplicity" cannot necessarily be determined by merely counting parameters. Viewed in one way, these results explain Occam's Razor, and point to the appropriate definition of simplicity. Viewed another way, however, they say that Bayesians needn't concern themselves with Occam's Razor, since to the extent that it is valid, it will be applied automatically anyway.

1.2 Bayesian neural networks

Workers in the field of "neural networks" have diverse backgrounds and motivations, some of which can be seen in the collection of Rumelhart and McClelland (1986b) and the books by Hertz, Krogh, and Palmer (1991), Bishop (1995), and Ripley (1996). In this book, I focus on the potential for neural networks to learn models for complex relationships that are interesting from the viewpoint of artificial intelligence or useful in engineering applications.

In statistical terms, neural networks are "nonparametric" models — a term meant to contrast them with simpler "parametric" models in which the relationship is characterized in terms of a few parameters, which often have meaningful interpretations. (The term "nonparametric" is somewhat of a misnomer in this context, however. These models do have parameters; they are just more numerous, and less interpretable, than those of "parametric" models.) Neural networks are not the only nonparametric models that can be applied to complex problems, of course, though they are among the more widely used such. I hope that the work on Bayesian learning for neural networks described in this book will ultimately be of help in devising and implementing other nonparametric Bayesian methods as well.

1.2.1 Multilayer perceptron networks

The neural networks most commonly used in engineering applications, and the only sort discussed in this book, are the *multilayer perceptron* networks (Rumelhart, Hinton, and Williams 1986a, 1986b), also known as "back-propagation" or "feedforward" networks. These networks take in a set of real inputs, x_i, and from them compute one or more output values, $f_k(x)$, perhaps using some number of layers of *hidden units*. In a typical network with one hidden layer, such as is illustrated in Figure 1.1, the outputs might be computed as follows:

$$f_k(x) = b_k + \sum_j v_{jk} h_j(x) \qquad (1.6)$$

$$h_j(x) = \tanh\left(a_j + \sum_i u_{ij} x_i\right) \qquad (1.7)$$

Here, u_{ij} is the *weight* on the connection from input unit i to hidden unit j; similarly, v_{jk} is the weight on the connection from hidden unit j to output

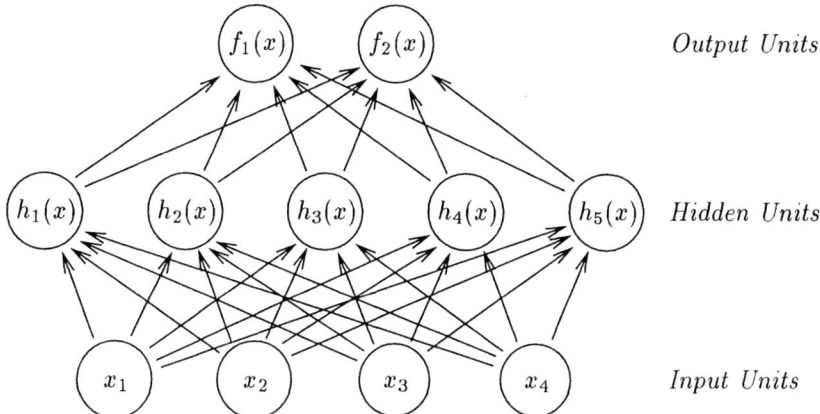

FIGURE 1.1. A multilayer perceptron with four input units, one layer of five hidden units, and two output units. The input units at the bottom are fixed to their values for a particular case. The values of the hidden units are then computed, followed by the values of the output units. The value for a hidden or output unit is a function of the weighted sum of values it receives from the units that are connected to it via the arrows.

unit k. The a_j and b_k are the *biases* of the hidden and output units. These weights and biases are the parameters of the network.

Each output value, $f_k(x)$, is just a weighted sum of hidden unit values, plus a bias. Each hidden unit computes a similar weighted sum of input values, and then passes it through a nonlinear *activation function*. The activation function chosen here is the hyperbolic tangent (tanh), an antisymmetric function of sigmoidal shape, whose value is close to -1 for large negative arguments, zero for a zero argument, and close to $+1$ for large positive arguments. A nonlinear activation function allows the hidden units to represent "hidden features" of the input that are useful in computing the appropriate outputs. If a linear activation function were used, the hidden layer could be eliminated, since equivalent results could be obtained using direct connections from the inputs to the outputs.

Several people (Cybenko 1989, Funahashi 1989, Hornik, Stinchcombe, and White 1989) have shown that a multilayer perceptron network with one hidden layer can approximate any function defined on a compact domain arbitrarily closely, if sufficient numbers of hidden units are used. Nevertheless, more elaborate network architectures may have advantages, and are commonly used. Possibilities include using more layers of hidden units, providing direct connections from inputs to outputs, and using different activation functions. However, in "feedforward" networks such as I consider here, the connections never form cycles, in order that the values of the outputs can be computed in a single forward pass, in time proportional to the number of network parameters.

Multilayer perceptron networks can be used to define probabilistic models for regression and classification tasks by using the network outputs to define the conditional distribution for one or more *targets*, y_k, given the various possible values of an input vector, x. The distribution of x itself is not modeled; it may not even be meaningful, since the input values might simply be chosen by the user. Models based on multilayer perceptrons have been applied to a great variety of problems. One typical class of applications are those that take as input sensory information of some type and from that predict some characteristic of what is sensed. Thodberg (1996), for example, predicts the fat content of meat from spectral information.

For a regression model with real-valued targets, the conditional distribution for the targets, y_k, given the input, x, might be defined to be Gaussian, with y_k having a mean of $f_k(x)$ and a standard deviation of σ_k. The different outputs are usually taken to be independent, given the input. We will then have

$$P(y \mid x) = \prod_k \frac{1}{\sqrt{2\pi}\sigma_k} \exp\bigl(-(f_k(x) - y_k)^2 / 2\sigma_k^2\bigr) \qquad (1.8)$$

The "noise levels", σ_k, might be fixed, or might be regarded as hyperparameters (which stretches the previously-given definition of this term, but corresponds to how these quantities are often treated).

For a classification task, where the target, y, is a single discrete value indicating one of K possible classes, the *softmax* model (Bridle 1989) can be used to define the conditional probabilities of the various classes using a network with K output units, as follows:

$$P(y = k \mid x) = \exp(f_k(x)) \bigg/ \sum_{k'} \exp(f_{k'}(x)) \qquad (1.9)$$

This method of defining class probabilities is also used in generalized linear models in statistics (McCullagh and Nelder, 1983, Section 5.1.3).

The weights and biases in neural networks are learned based on a set of *training cases*, $(x^{(1)}, y^{(1)}), \ldots, (x^{(n)}, y^{(n)})$, giving examples of inputs, $x^{(i)}$, and associated targets, $y^{(i)}$ (both of which may have several components). Standard neural network training procedures adjust the weights and biases in the network so as to minimize a measure of "error" on the training cases, most commonly, the sum of the squared differences between the network outputs and the targets. Minimization of this error measure is equivalent to maximum likelihood estimation for the Gaussian noise model of equation (1.8), since minus the log of the likelihood with this model is proportional to the sum of the squared errors.

Finding the weights and biases that minimize the chosen error function is commonly done using some gradient-based optimization method, using derivatives of the error with respect to the weights and biases that are calculated by *backpropagation* (Rumelhart, Hinton, and Williams 1986a, 1986b). There are typically many local minima, but good solutions are often found despite this.

To reduce overfitting, a penalty term proportional to the sum of the squares of the weights and biases is often added to the error function, resulting in a maximum penalized likelihood estimation procedure. This modification is known as *weight decay*, because its effect is to bias the procedure in favour of small weights. Determining the proper magnitude of the weight penalty is difficult — with too little weight decay, the network may "overfit", but with too much weight decay, the network will "underfit", ignoring the data.

The method of *cross validation* (Stone 1974) is sometimes used to find an appropriate weight penalty. In the simplest form of cross validation, the amount of weight decay is chosen to optimize performance on a validation set separate from the cases used to estimate the network parameters. This method does not make efficient use of the available training data, however. In n-way cross validation, the training set is partitioned into n subsets, each of which is used as the validation set for a network trained on the other $n-1$ subsets. Total error on all these validation sets is used to pick a good amount of weight decay, which is then used in training a final network on all the data. This procedure is computationally expensive, however, and could run into problems if the n networks find dissimilar local minima, for which different weight penalties are appropriate.

In the Bayesian approach to neural network learning, the objective is to find the predictive distribution for the target values in a new "test" case, given the inputs for that case, and the inputs and targets in the training cases. Since the distribution of the inputs is not being modeled, the predictive distribution of equation (1.4) is modified as follows:

$$P(y^{(n+1)} \mid x^{(n+1)}, (x^{(1)}, y^{(1)}), \ldots, (x^{(n)}, y^{(n)}))$$
$$= \int P(y^{(n+1)} \mid x^{(n+1)}, \theta) \, P(\theta \mid (x^{(1)}, y^{(1)}), \ldots, (x^{(n)}, y^{(n)})) \, d\theta \quad (1.10)$$

Here, θ represents the network parameters (weights and biases). The posterior density for these parameters is proportional to the product of whatever prior is being used and the likelihood function, as in equation (1.3). The likelihood is slightly modified because the distribution of the inputs is not being modeled:

$$L(\theta \mid (x^{(1)}, y^{(1)}), \ldots, (x^{(n)}, y^{(n)})) = \prod_{i=1}^{n} P(y^{(i)} \mid x^{(i)}, \theta) \quad (1.11)$$

The distribution for the target values, $y^{(i)}$, given the corresponding inputs, $x^{(i)}$, and the parameters of the network is defined by the type of model with which the network is being used; for regression and softmax classification models it is given by equations (1.8) and (1.9).

If we wish to guess a component of $y^{(n+1)}$, with squared error loss, the best strategy is to guess the mean of its predictive distribution. For a regression model, this reduces to the following guess:

$$\widehat{y}_k^{(n+1)} = \int f_k(x^{(n+1)}, \theta) \, P(\theta \mid (x^{(1)}, y^{(1)}), \ldots, (x^{(n)}, y^{(n)})) \, d\theta \quad (1.12)$$

Here the network output functions, f_k, are written with the dependence on the network parameters, θ, being shown explicitly.

1.2.2 Selecting a network model and prior

At first sight, the Bayesian framework may not appear suitable for use with neural networks. Bayesian inference starts with a prior for the model parameters, which is supposed to embody our prior beliefs about the problem. In a multilayer perceptron network, the parameters are the connection weights and unit biases, whose relationship to anything that we might know about the problem seems obscure. The Bayesian engine thus threatens to stall at the outset for lack of a suitable prior.

However, to hesitate because of such qualms would be contrary to the spirit of the neural network field. MacKay (1991, 1992b) has tried the most obvious possibility of giving the weights and biases Gaussian prior distributions. This turns out to produce results that are at least reasonable. In his work, MacKay emphasizes the advantages of hierarchical models. He gives results of Bayesian learning for a network with one hidden layer, applied to a regression problem, in which he lets the variance of the Gaussian prior for the weights and biases be a hyperparameter. This allows the model to adapt to whatever degree of smoothness is indicated by the data. Indeed, MacKay discovers that the results are improved by using several variance hyperparameters, one for each type of parameter (weights out of input units, biases of hidden units, and weights and biases of output units). He notes that this makes sense in terms of prior beliefs if the inputs and outputs of the network are quantities of different sorts, measured on different scales, since in this case the effect of using a single variance hyperparameter would depend on the arbitrary choice of measurement units.

In a Bayesian model of this type, the role of the hyperparameters controlling the priors for weights is roughly analogous to the role of a weight decay constant in conventional training. With Bayesian training, values for these hyperparameters (more precisely, a distribution of values) can be found without the need for a validation set.

Buntine and Weigend (1991) discuss several possible schemes for prior distributions, such as priors that favour networks that produce high or low entropy predictions, or that compute smooth functions. The degree of preference imposed can be controlled by a hyperparameter. Their treatment of smoothness priors applies only to simple networks, however. This work links the choice of prior for weights to the actual effects of these weights on the function computed by the network, which is clearly necessary if we are to choose a prior that represents our beliefs about this function.

This past work shows that useful criteria for selecting a suitable prior can sometimes be found even without a full understanding of what the priors over weights and biases mean in terms of the functions computed by the network. Still, the selection of a particular network architecture and associated prior remains *ad hoc*. Bayesian neural network users may have difficulty claiming with a straight face that their models and priors are selected because they are just what is needed to capture their prior beliefs about the problem.

The work I describe in Chapter 2 addresses this problem. Applying the philosophy of Bayesian learning for complex problems outlined in Section 1.1.4, I focus on priors for networks with an infinite number of hidden units. (In practice, such networks would be approximated by large finite networks.) Use of an infinite network is in accord with prior beliefs, since seldom will we believe that the true function we are learning can be exactly represented by any finite network. In addition, the characteristics of priors for infinite networks can often be found analytically. Further insight into the nature of these priors can be obtained by randomly generating networks from the prior and visually examining the functions that these networks compute. In Chapter 4, I report the results of applying networks with relatively large numbers of hidden units to actual data sets.

1.2.3 Automatic Relevance Determination (ARD) models

Another dimension of complexity in neural network models is the number of input variables used in modeling the distribution of the targets. In many problems, there will be a large number of potentially measurable attributes which could be included as inputs if we thought this would improve predictive performance. Unlike the situation with respect to hidden units, however, including more and more inputs (all on an equal footing) must ultimately lead to poor performance, since with enough inputs, it is inevitable that an input which is in fact irrelevant will by chance appear in a finite training set to be more closely associated with the targets than are the truly relevant inputs. Predictive performance on test cases will then be poor.

Accordingly, we must limit the number of input variables we use, based on our assessment of which attributes are most likely to be relevant. (Alternatively, if we do include a huge number of inputs that we think are probably irrelevant, we must use an asymmetrical prior that expresses our belief that some inputs are less likely to be relevant than others.) However, in problems where the underlying mechanisms are not well understood, we will not be confident as to which are the relevant attributes. The inputs we choose to include will be those that we believe may *possibly* be relevant, but we will also believe that some of these inputs may turn out to have little or no relevance. We would therefore like to use models that can automatically determine the degree to which such inputs of unknown relevance are in fact relevant.

Models of this sort have been developed by David MacKay and myself, and used by MacKay in a model of energy usage in buildings (Mackay 1994a). In such an *Automatic Relevance Determination (ARD)* model, each input variable has associated with it a hyperparameter that controls the magnitudes of the weights on connections out of that input unit. These hyperparameters are given some prior distribution, and conditional on the values of these hyperparameters, the weights out of each input have independent Gaussian prior distributions with standard deviation given by the corresponding hyperparameter. If the hyperparameter associated with an input specifies a small standard deviation for weights out of that input, these weights will likely all be small, and the input will have little effect on the output; if the hyperparameter specifies a large standard deviation, the effect of the input will likely be significant. The posterior distributions of these hyperparameters will reflect which of these situations is more probable, in light of the training data.

ARD models are intended for use with a complex network in which each input is associated with many weights, with the role of the ARD hyperparameters being to introduce dependencies between these weights. In such a situation, if the weight on one connection out of an input becomes large, indicating that the input has some relevance, this will influence the distribution of the associated hyperparameter, which in turn will make it more likely that other weights out of the same input will also be large.

Formally, one could define an ARD model for a network with a single target and no hidden units, in which each input unit connects only to the target (a network equivalent to a simple linear regression model). However, each ARD hyperparameter in this simple network would control the distribution of only a single weight, eliminating its role in introducing dependencies. By integrating over the ARD hyperparameters, we could produce a direct specification for the prior over weights in which each weight would be independent of the others, but would now have some prior distribution other than a Gaussian. This might or might not be a good model, but in either case, it seems likely that its properties could be more easily understood in this direct formulation, with the hyperparameters eliminated. On the other hand, this method of obtaining a non-Gaussian prior might have computational advantages in some contexts.

Although use of ARD models may seem to be straightforward extension of MacKay's previous use of several hyperparameters to control the distribution of different classes of weights (see Section 1.2.2), these models in fact raise several subtle issues. Just what do we mean by a "large" or "small" value of the standard deviation for the prior over weights associated with a particular input? The answer must depend somehow on the measurement units used for this input. What prior should we use for the ARD hyperparameters? It would be convenient if we could use a vague prior, but it is not clear that this will give the best results. These issues are discussed further in Chapter 4, where ARD models are evaluated on several data sets.

1.2.4 An illustration of Bayesian learning for a neural net

An example will illustrate the general concept of Bayesian learning, its application to neural networks, and the infeasibility of brute force methods of Bayesian computation for problems of significant size.

Figure 1.2 shows Bayesian learning in action for a regression model based on a neural network with one input, one output, and 16 hidden units. The operation of the network is described by equations (1.6) and (1.7). The conditional distribution for the target is given by equation (1.8), with the noise level set to $\sigma = 0.1$.

On the left of the figure are the functions computed by ten such networks whose weights and biases were drawn from independent Gaussian prior distributions, each with mean zero and standard deviation one, except for the output weights, which had standard deviation $1/\sqrt{16}$. As explained in Chapter 2, setting the standard deviation of the output weights to be inversely proportional to the square root of the number of hidden units ensures that the prior over functions computed by the network reaches a sensible limit as the number of hidden units goes to infinity.

On the right of Figure 1.2 are ten functions drawn from the posterior distribution that results when this prior is combined with the likelihood due to the six data points shown (see equations (1.3) and (1.11)). As one would hope, the posterior distribution is concentrated on functions that pass near the data points.

The best way to guess the targets associated with various input values, assuming we wish to minimize the expected squared error in the guesses, is to use the average of the network functions over the posterior distribution of network parameters (as in equation (1.12)). We can make a Monte Carlo estimate of this average across the posterior by averaging the ten functions

18   Chapter 1. Introduction

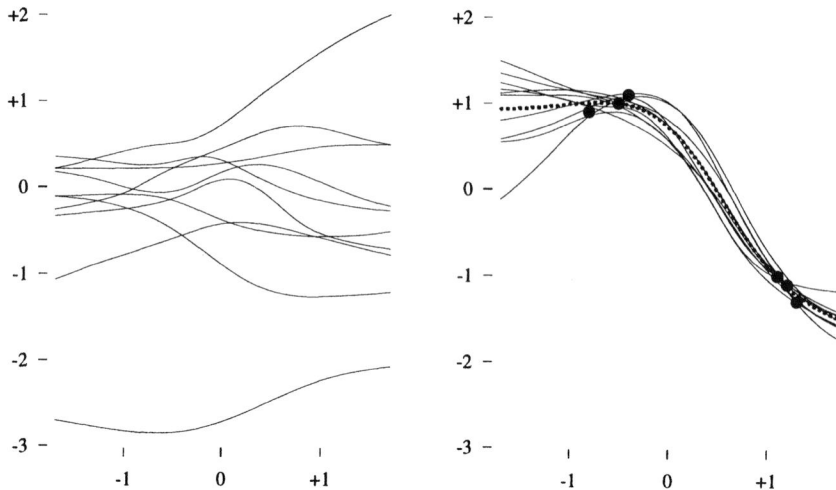

FIGURE 1.2. An illustration of Bayesian inference for a neural network. On the left are the functions computed by ten networks whose weights and biases were drawn at random from Gaussian prior distributions. On the right are six data points and the functions computed by ten networks drawn from the posterior distribution derived from the prior and the likelihood due to these data points. The heavy dotted line is the average of the ten functions drawn from the posterior, which is an approximation to the function that should be guessed in order to minimize expected squared error loss.

shown that were drawn from the posterior. This averaged function is shown in the figure by a heavy dotted line. Bayesian inference provides more than just a single-valued guess, however. By examining the sample of functions from the posterior, we can also see how uncertain these guesses are. We can, for example, see that the uncertainty increases rapidly beyond the region where the training points are located.

Figure 1.2 was produced using a simple algorithm that is of interest both because it illuminates the nature of Bayesian learning, and because it illustrates that direct approaches to performing Bayesian inference can rapidly become infeasible as the problem becomes bigger.

The left half of the figure was easy to produce, since generating values for the network weights and biases from independent Gaussian distributions can be done quickly using standard methods (Devroye 1986). It is, in fact, very often the case that sampling from the prior is simple and fast, even for complex models.

The right half of the figure was produced by generating many networks from the prior, computing the likelihood for each based on the six training points, and then accepting each network with a probability proportional to its likelihood, with the constant of proportionality chosen to make the maximum probability of acceptance be one. Networks that were not accepted were discarded, with the process continuing until ten networks had been accepted; these ten are shown in the figure.

This algorithm — a form of *rejection sampling* (Devroye 1986) — directly embodies the definition of the posterior given by equation (1.3). The prior contributes to the result by controlling the generation of candidate networks; the likelihood contributes by controlling which of these candidates are accepted. The algorithm is not very efficient, however. As can be seen by looking at the right of Figure 1.2, the functions computed by most networks drawn from the prior do not pass near the training points (within a few standard deviations, with $\sigma = 0.1$) — in fact, none of the ten functions shown there are close to all the data points. The number of functions that will have to be drawn from the prior before one is accepted will therefore be high. Generating the sample of ten functions from the posterior shown in the figure turned out to require generating 2.6 million networks from the prior.

As the number of data points in the training set increases, the time required by this method grows exponentially. More efficient methods are clearly needed in practice.

1.2.5 Implementations based on Gaussian approximations

The posterior distribution for the parameters (weights and biases) of a multilayer perceptron network is typically very complex, with many modes. Finding the predictive distribution for a test case by evaluating the integral of equation (1.10) is therefore a difficult task. In Chapter 3, I address this problem using Markov chain Monte Carlo methods. Here, I will discuss implementations based on Gaussian approximations to modes, which have been described by Buntine and Weigend (1991), MacKay (1991, 1992b, 1992c), and Thodberg (1996). Hinton and van Camp (1993) use a Gaussian approximation of a different sort.

Schemes based on Gaussian approximations to modes operate as follows:

1) Find one or more modes of the posterior parameter distribution.

2) Approximate the posterior distribution in the vicinity of each such mode by a Gaussian whose inverse covariance matrix matches the second derivatives of the log posterior at the mode.

3) If more than one mode is being used, decide how much weight to give to each.

4) Approximate the predictive distribution of equation (1.10) by the corresponding integral with respect to the Gaus-

sian approximation to the mode, or the weighted mixture of Gaussians approximating the several modes.

Step (4) is easy for models that are linear in the vicinity of a mode. Simple approximations may suffice in other cases (MacKay 1992c). At worst, it can be done fairly efficiently by simple Monte Carlo methods (Ripley 1994a).

I have not mentioned above how to handle hyperparameters, such as the prior variances for groups of weights, and the noise level for a regression problem. This is a matter about which there has been some controversy.

Buntine and Weigend (1991) analytically integrate over the hyperparameters, and then look for modes of the resulting marginal posterior distribution for the parameters. Eliminating the hyperparameters in this way may appear to be an obviously beneficial simplification of the problem, but this is not the case — as MacKay (1994b) explains, integrating out such hyperparameters can sometimes produce a marginal posterior parameter distribution in which the mode is entirely unrepresentative of the distribution as a whole. Basing an approximation on the location of the mode will then give drastically incorrect results.

In MacKay's implementation (1991, 1992b, 1992c), he assumes only that the Gaussian approximation can be used to represent the posterior distribution of the parameters for given values of the hyperparameters. He fixes the hyperparameters to the values that maximize the probability of the data (what he calls the "evidence" for these values of the hyperparameters). In finding these values, he makes use of the Gaussian approximation to integrate over the network parameters.

MacKay's "evidence" approach to handling the hyperparameters is computationally equivalent to the "ML-II" method of prior selection (Berger 1982, Section 3.5.4). From a fully Bayesian viewpoint, it is only an approximation to the true answer, which would be obtained by integrating over the hyperparameters as well as the parameters, but experience has shown that it is often a good approximation. Wolpert (1993) criticizes the use of this procedure for neural networks on the grounds that by analytically integrating over the hyperparameters, in the manner of Buntine and Weigend, one can obtain the relative posterior probability densities for different values of the network parameters *exactly*, without the need for any approximation. This criticism is based on a failure to appreciate the nature of the task. The posterior probability densities for different parameter values are, in themselves, of no interest — all that matters is how well the predictive distribution is approximated. MacKay (1994b) shows that in approximating this predictive distribution, it is more important to integrate over the large number of parameters in the network than over the typically small number of hyperparameters.

This controversy has perhaps distracted attention from other problems with Gaussian approximation methods that I believe are more significant.

First, how should one handle the presence of multiple modes? One approach is to ignore the problem, simply assuming that all the modes are about equally good. The general success of neural network learning procedures despite the presence of local minima suggests that this approach may not be as ridiculous as it might sound. Nevertheless, one would like to do better, finding several modes, and making predictions based on a weighted average of the predictions from each mode. One possibility is to weight each mode by an estimate of the total probability mass in its vicinity, obtained from the relative probability density at the mode and the determinant of the covariance matrix of the Gaussian used to approximate the mode (Buntine and Weigend 1991, Ripley 1994a). This is not a fully correct procedure, however — the weight a mode receives ought really to be adjusted according to the probability of the mode being found by the optimization procedure, with the easily found modes being given less weight than they would otherwise have had, since they occur more often. For large problems this will not be possible, however, since each mode will typically be seen only once, making the probabilities of finding the modes impossible to determine. Another problem is that if the Gaussian approximation is not very accurate, one mode may receive most of the weight simply because it happened to be favoured by approximation error. Such problems lead Thodberg (1996) to use the estimated probability mass only to select a "committee" based on the better modes (perhaps from different models), to each of which he assigns equal weight.

A second, potentially more serious, question is whether the Gaussian approximation for the distribution in the vicinity of a mode is reasonably good (even for fixed values of the hyperparameters). One reason for optimism in this regard is that the posterior distribution for many models becomes increasingly Gaussian as the amount of training data increases (DeGroot 1970, Chapter 10). However, if we subscribe to the view of complexity presented in Section 1.1.4, we should not confine ourselves to simple models, for which this asymptotic result may be relevant, but should instead use as complex a model as we can handle computationally, in order to extract the maximum information from the data, and obtain a full indication of the remaining uncertainty. I believe that the Gaussian approximation will seldom be good for such complex models.

Looking at neural network models in particular, the following argument suggests that the Gaussian approximation may be bad when the amount of data is insufficient to determine the values of the weights out of the hidden units, to within a fairly small fraction of their values. In a multivariate Gaussian, the conditional distribution of one variable given values for the other variables has a variance that is independent of the particular values the other variables take (these affect only the conditional mean). Accordingly, for the Gaussian approximation to the posterior distribution of the weights in a network to be good, the conditional distribution for a weight into a hidden unit must have a variance almost independent of the values of the weights out of that hidden unit. Since the weights out of a hidden unit have a multiplicative effect on the hidden unit's influence, this can be true only if the posterior variation in these weights is small compared to their magnitude.

As will be seen in Chapter 2, when reasonable priors are used, all or most of the weights out of the hidden units in a large network will be small, and, individually, each such hidden unit will have only a small influence on the network output. In the posterior distribution, the variation in the weights out of these hidden units will thus be large compared to their magnitudes, and we should not expect the Gaussian approximation to work well.

Finally, Hinton and van Camp (1993) take a rather different approach to approximating the posterior weight distribution by a Gaussian. They employ an elaboration of the Minimum Description Length framework (Rissanen 1986) that is equivalent to Bayesian inference using an approximation to the posterior distribution chosen so as to minimize the Kullback-Leibler divergence with the true posterior. Hinton and van Camp choose to approximate the posterior by a Gaussian with a diagonal covariance matrix. Note that the Gaussian of this class that minimizes the Kullback-Leibler divergence with the true posterior will not necessarily be positioned at a mode (though one might expect it to be close). For the reasons just outlined, we may expect that Gaussian approximations of this sort will also fail to be good for large networks in which the weights are not well determined.

1.3 Markov chain Monte Carlo methods

In Chapter 3, I will present an implementation of Bayesian learning for neural networks in which the difficult integrations required to make predictions are performed using Markov chain Monte Carlo methods. These methods have been used for many years to solve problems in statistical physics, and have recently been widely applied to Bayesian models in statistics. Markov chain Monte Carlo methods make no assumptions concerning the form of the distribution, such as whether it can be approximated by a Gaussian. In theory at least, they take proper account of multiple modes, as well as the possibility that the dominant contribution to the integral may come from areas not in the vicinity of any mode. The main disadvantage of Markov chain methods is that they may in some circumstances require a very long time to converge to the desired distribution.

The implementation in Chapter 3 is based on the "hybrid Monte Carlo" algorithm, which was developed for applications in quantum chromodynamics, and has not previously been applied in a statistical context. In this section, I describe the basic concept of Markov chain Monte Carlo, and review two better-known methods on which the hybrid Monte Carlo algorithm is based; I leave the exposition of the hybrid Monte Carlo algorithm itself to Chapter 3. I have reviewed these methods in more detail elsewhere (Neal 1993b). Tierney (1994) and Smith and Roberts (1993) also review recent work on Markov chain Monte Carlo methods and their applications in statistics.

1.3.1 Monte Carlo integration using Markov chains

The objective of Bayesian learning is to produce predictions for test cases. This may take the form of finding predictive probabilities, as in equation (1.10), or of making single-valued guesses, as in equation (1.12). Both tasks require that we evaluate the expectation of a function with respect to the posterior distribution for model parameters. Writing the posterior probability density for the parameters as $Q(\theta)$, the expectation of $a(\theta)$ is

$$E[a] \;=\; \int a(\theta)\, Q(\theta)\, d\theta \qquad (1.13)$$

For example, by letting $a(\theta) = f_k(x^{(n+1)}, \theta)$, we get the integral of equation (1.12), used to find the best guess for $y_k^{(n+1)}$ under squared error loss.

Such expectations can be estimated by the *Monte Carlo* method, using a sample of values from Q:

$$E[a] \;\approx\; \frac{1}{N} \sum_{t=1}^{N} a(\theta^{(t)}) \qquad (1.14)$$

where $\theta^{(1)}, \ldots, \theta^{(N)}$ are generated by a process that results in each of them having the distribution defined by Q. In simple Monte Carlo methods, the $\theta^{(t)}$ are independent. When Q is a complicated distribution, generating such independent values is often infeasible, but it may nevertheless be possible to generate a series of dependent values. The Monte Carlo integration formula of equation (1.14) still gives an unbiased estimate of $E[a]$ even when the $\theta^{(t)}$ are dependent, and as long as the dependence is not too great, the estimate will still converge to the true value as N increases.

Such a series of dependent values may be generated using a *Markov chain* that has Q as its stationary distribution. The chain is defined by giving an *initial distribution* for the first state of the chain, $\theta^{(1)}$, and a set of *transition probabilities* (or densities) for a new state, $\theta^{(t+1)}$, to follow the current state, $\theta^{(t)}$. The probability densities for these transitions will be written as $T(\theta^{(t+1)} \mid \theta^{(t)})$. An *invariant* (or *stationary*) distribution, Q, is one that persists once it is established — that is, if $\theta^{(t)}$ has the distribution given by Q, then $\theta^{(t')}$ will have the same distribution for all $t' > t$. This invariance condition can be written as follows:

$$Q(\theta') = \int T(\theta' \mid \theta) \, Q(\theta) \, d\theta \qquad (1.15)$$

Invariance with respect to Q is implied by the stronger condition of *detailed balance* — that for all θ and θ':

$$T(\theta' \mid \theta) \, Q(\theta) = T(\theta \mid \theta') \, Q(\theta') \qquad (1.16)$$

A chain satisfying detailed balance is said to be *reversible*.

A Markov chain that is *ergodic* has a unique invariant distribution, its *equilibrium* distribution, to which it converges from any initial state. If we can find an ergodic Markov chain that has Q as its equilibrium distribution, we can estimate expectations with respect to Q using equation (1.14), with $\theta^{(1)}, \ldots, \theta^{(N)}$ being the states of the chain, perhaps with some early states discarded, since they may not be representative of the equilibrium distribution. Because of the dependencies between the $\theta^{(t)}$, the number of values for θ needed for the Monte Carlo estimate to reach a certain level of accuracy may be larger than would be required if the $\theta^{(t)}$ were independent, sometimes much larger. The chain may also require a long time to reach a point where the distribution of the current state is a good approximation to the equilibrium distribution.

The effect of dependencies on the accuracy of a Monte Carlo estimate can be quantified in terms of the *autocorrelations* between the values of $a(\theta^{(t)})$ once equilibrium has been reached (see, for example, (Ripley 1987, Neal 1993b)). If a has finite variance, the variance of the estimate of $E[a]$ given by equation 1.14 will be $\mathrm{Var}[a]/N$ if the $\theta^{(t)}$ are independent. When the $\theta^{(t)}$ are dependent, and N is large, the variance of the estimate is $\mathrm{Var}[a] / (N/\tau)$, where

$$\tau = 1 + 2 \sum_{s=1}^{\infty} \rho(s) \qquad (1.17)$$

is a measure of the inefficiency due to the presence of dependencies. Here, $\rho(s)$ is the autocorrelation of a at lag s, defined by

$$\rho(s) = \frac{E\bigl[(a(\theta^{(t)}) - E[a])\,(a(\theta^{(t-s)}) - E[a])\bigr]}{\mathrm{Var}[a]} \qquad (1.18)$$

Since we assume that equilibrium has been reached, the value used for t does not affect the above definition. For Markov chains used to sample complex distributions, these autocorrelations are typically positive, leading to a value for τ greater than one. (It is possible for τ to be less than one, however, in which case the dependencies actually increase the accuracy of the estimate.)

To use the Markov chain Monte Carlo method to estimate an expectation with respect to some distribution, Q, we need to construct a Markov chain which is ergodic, which has Q as its equilibrium distribution, which converges to this distribution as rapidly as possible, and in which the states visited once the equilibrium distribution is reached are not highly dependent. To construct such a chain for a complex problem, we can combine the transitions for simpler chains, since as long as each such transition leaves Q invariant, the result of applying these transitions in sequence will also leave Q invariant. In the remainder of this section, I will review two simple methods for constructing Markov chains that will form the basis for the implementation described in Chapter 3.

1.3.2 Gibbs sampling

Gibbs sampling, known in the physics literature as the *heatbath* method, is perhaps the simplest Markov chain Monte Carlo method. It is used in the "Boltzmann machine" neural network of Ackley, Hinton, and Sejnowski (1985) to sample from distributions over stochastic hidden units, and has become widely used for statistical problems, following its exposition by Geman and Geman (1984) and by Gelfand and Smith (1990).

Gibbs sampling is applicable when we wish to sample from a distribution over a multi-dimensional parameter, $\theta = \{\theta_1, \ldots, \theta_p\}$. Presumably, directly sampling from the distribution given by $Q(\theta)$ is infeasible, but we assume that we can generate a value from the conditional distribution (under Q) for one component of θ given values for all the other components of θ. This allows us to simulate a Markov chain in which $\theta^{(t+1)}$ is generated from $\theta^{(t)}$ as follows:

Pick $\theta_1^{(t+1)}$ from the distribution of θ_1 given $\theta_2^{(t)}, \theta_3^{(t)}, \ldots, \theta_p^{(t)}$

Pick $\theta_2^{(t+1)}$ from the distribution of θ_2 given $\theta_1^{(t+1)}, \theta_3^{(t)}, \ldots, \theta_p^{(t)}$

$$\vdots$$

Pick $\theta_j^{(t+1)}$ from the distribution of θ_j given $\theta_1^{(t+1)}, \ldots, \theta_{j-1}^{(t+1)}, \theta_{j+1}^{(t)}, \ldots, \theta_p^{(t)}$

$$\vdots$$

Pick $\theta_p^{(t+1)}$ from the distribution of θ_p given $\theta_1^{(t+1)}, \theta_2^{(t+1)}, \ldots, \theta_{p-1}^{(t+1)}$

Note that the new value for θ_j is used immediately when picking a new value for θ_{j+1}.

Such transitions will leave the desired distribution, Q, invariant if all the steps making up each transition leave Q invariant. Since step j leaves θ_k for

$k \neq j$ unchanged, the desired marginal distribution for these components is certainly invariant. Furthermore, the conditional distribution for θ_j in the new state given the other components is defined to be that which is desired. Together, these ensure that if we started from the desired distribution, the joint distribution for all the θ_j after all the above steps must also be the desired distribution. These transitions do not necessarily lead to an ergodic Markov chain, however; this must be established in each application.

Whether Gibbs sampling is useful for Bayesian inference depends on whether the posterior distribution of one parameter conditional on given values for the other parameters can easily be sampled from. For many statistical problems, these conditional distributions are of standard forms for which efficient generation procedures are known. For neural networks, however, the posterior conditional distribution for one weight in the network given values for the other weights can be extremely messy, with many modes. There appears to be no reasonable way of applying Gibbs sampling in this case. However, Gibbs sampling is one component of the hybrid Monte Carlo algorithm, which can be used for neural networks. In the implementation of Chapter 3, it will also be used to update hyperparameters.

1.3.3 The Metropolis algorithm

The *Metropolis algorithm* was introduced in the classic paper of Metropolis, Rosenbluth, Rosenbluth, Teller, and Teller (1953), and has since seen extensive use in statistical physics. It is also the basis for the widely-used optimization method of "simulated annealing" (Kirkpatrick, Gelatt, and Vecchi 1983).

In the Markov chain defined by the Metropolis algorithm, a new state, $\theta^{(t+1)}$, is generated from the previous state, $\theta^{(t)}$, by first generating a *candidate state* using a specified *proposal distribution*, and then deciding whether or not to accept the candidate state, based on its probability density relative to that of the old state, with respect to the desired invariant distribution, Q. If the candidate state is accepted, it becomes the next state of the Markov chain; if the candidate state is instead rejected, the new state is the same as the old state, and is included again in any averages used to estimate expectations.

In detail, the transition from $\theta^{(t)}$ to $\theta^{(t+1)}$ is defined as follows:

1) Generate a candidate state, θ^*, from a proposal distribution that may depend on the current state, with density given by $S(\theta^* \mid \theta^{(t)})$.

2) If $Q(\theta^*) \geq Q(\theta^{(t)})$, accept the candidate state; otherwise, accept the candidate state with probability $Q(\theta^*)/Q(\theta^{(t)})$.

3) If the candidate state is accepted, let $\theta^{(t+1)} = \theta^*$; if the candidate state is rejected, let $\theta^{(t+1)} = \theta^{(t)}$.

The proposal distribution must be symmetrical, satisfying the condition $S(\theta' \mid \theta) = S(\theta \mid \theta')$. In some contexts, $Q(\theta)$ is defined in terms of an "energy" function, $E(\theta)$, with $Q(\theta) \propto \exp(-E(\theta))$. In step (2), one then always accepts candidate states with lower energy, and accepts states of higher energy with probability $\exp(-(E(\theta^*) - E(\theta^{(t)})))$.

To show that these transitions leave Q invariant, we first need to write down the transition probability density function. This density function is singular, however, since there is a non-zero probability that the new state will be exactly the same as the old state. Fortunately, in verifying the detailed balance condition (equation (1.16)), we need pay attention only to transitions that change the state. For $\theta' \neq \theta$, the procedure above leads to the following transition densities:

$$T(\theta' \mid \theta) = S(\theta' \mid \theta) \min\left(1, Q(\theta')/Q(\theta)\right) \tag{1.19}$$

Detailed balance can thus be verified as follows:

$$\begin{aligned} T(\theta' \mid \theta)\, Q(\theta) &= S(\theta' \mid \theta) \min\left(1, Q(\theta')/Q(\theta)\right) Q(\theta) & (1.20) \\ &= S(\theta' \mid \theta) \min\left(Q(\theta), Q(\theta')\right) & (1.21) \\ &= S(\theta \mid \theta') \min\left(Q(\theta'), Q(\theta)\right) & (1.22) \\ &= S(\theta \mid \theta') \min\left(1, Q(\theta)/Q(\theta')\right) Q(\theta') & (1.23) \\ &= T(\theta \mid \theta')\, Q(\theta') & (1.24) \end{aligned}$$

The Metropolis updates therefore leave Q invariant. Note, however, that they do not always produce an ergodic Markov chain; this depends on details of Q, and on the proposal distribution used.

Many choices are possible for the proposal distribution of the Metropolis algorithm. One simple choice is a Gaussian distribution centred on $\theta^{(t)}$, with standard deviation chosen so that the probability of the candidate state being accepted is reasonably high. (A very low acceptance rate is usually bad, since successive states are then highly dependent.) When sampling from a complex, high-dimensional distribution, the standard deviation of such a proposal distribution will often have to be small, compared to the extent of Q, since large changes will almost certainly lead to a region of low probability. This will result in a high degree of dependence between successive states, since many steps will be needed to move to a distant point in the distribution. This problem is exacerbated by the fact that these movements take the form of a random walk, rather than a systematic traversal.

Due to this problem, simple forms of the Metropolis algorithm can be very slow when applied to problems such as Bayesian learning for neural

28   Chapter 1. Introduction

networks. As will be seen in Chapter 3, however, this problem can be alleviated by using the hybrid Monte Carlo algorithm, in which candidate states are generated by a dynamical method that largely avoids the random walk aspect of the exploration.

1.4   Outline of the remainder of the book

The main part of this book deals with three issues concerning Bayesian learning for neural networks.

In Chapter 2, I examine the properties of prior distributions for neural networks, focusing on the limit as the number of hidden units in the network goes to infinity. My aim is to show that reasonable priors for such infinite networks can be defined, and to develop an understanding of the properties of such priors, so that we can select an appropriate prior for a particular problem.

In Chapter 3, I address the computational problem of producing predictions based on Bayesian neural network models. Such predictions involve integrations over the posterior distribution of network parameters (see equation (1.10)), which I estimate using a Markov chain Monte Carlo method based on the hybrid Monte Carlo algorithm. The aim of this work is to produce the predictions mathematically implied by the model and prior being used, using a feasible amount of computation time.

In Chapter 4, I evaluate how good the predictions of Bayesian neural network models are, using the implementation of Chapter 3. One of my aims is to further demonstrate that Bayesian inference does not require limiting the complexity of the model based on the amount of training data, as was already shown in Chapter 2. I also evaluate the effectiveness of hierarchical models, in particular the Automatic Relevance Determination model. The tests on real data sets demonstrate that the Bayesian approach, implemented using hybrid Monte Carlo, can be effectively applied to problems of moderate size.

Finally, in Chapter 5, I summarize the contributions of this work, and describe further work done by myself and others since the completion of the original thesis on which this book is based. I also indicate possible directions for future research.

Chapter 2

Priors for Infinite Networks

In this chapter, I show that priors over network parameters can be defined in such a way that the corresponding priors over functions computed by the network reach reasonable limits as the number of hidden units goes to infinity. When using such priors, there is thus no need to limit the size of the network in order to avoid "overfitting". The infinite network limit also provides insight into the properties of different priors. A Gaussian prior for hidden-to-output weights results in a Gaussian process prior for functions, which may be smooth, Brownian, or fractional Brownian. Quite different effects can be obtained using priors based on non-Gaussian stable distributions. In networks with more than one hidden layer, a combination of Gaussian and non-Gaussian priors appears most interesting.

The starting point for Bayesian inference is a prior distribution over the model parameters, which for a multilayer perceptron ("backprop") network are the connection weights and unit biases. This prior distribution is meant to capture our prior beliefs about the relationship we are modeling. When training data is obtained, the prior is updated to a posterior parameter distribution, which is then used to make predictions for test cases.

A problem with this approach is that the meaning of the weights and biases in a neural network is obscure, making it hard to design a prior distribution that expresses our beliefs. Furthermore, a network with a small number of hidden units can represent only a limited set of functions, which will generally not include the true function. Hence our actual prior belief will usually be that the model is simply wrong.

30   Chapter 2. Priors for Infinite Networks

I propose to address these problems by focusing on the limit as the number of hidden units in the network approaches infinity. Several people (Cybenko 1989, Funahashi 1989, Hornik, Stinchcombe, and White 1989) have shown that in this limit a multilayer perceptron network with one layer of hidden units can approximate any continuous function defined on a compact domain arbitrarily closely. An infinite network will thus be a reasonable "nonparametric" model for many problems. Furthermore, it turns out that in the infinite network limit we can easily analyse the nature of the priors over functions that result when we use certain priors for the network parameters. This allows us to select an appropriate prior based on our knowledge of the characteristics of the problem, or to set up a hierarchical model in which these characteristics can be inferred from the data.

In practice, of course, we will have to use networks with only a finite number of hidden units. The hope is that our computational resources will allow us to train a network of sufficient size that its characteristics are close to those of an infinite network.

Note that in this approach one does *not* restrict the size of the network based on the size of the training set — rather, the only limiting factors are the size of the computer used and the time available for training. Experience training networks by methods such as maximum likelihood might lead one to expect a large network to "overfit" a small training set, and perform poorly on later test cases. This does not occur with Bayesian learning, provided the width of the prior used for hidden-to-output weights is scaled down in a simple fashion as the number of hidden units increases, as required for the prior to reach a limit.

These statements presume that the implementation of Bayesian inference used produces the mathematically correct result. Achieving this is not trivial. Methods based on making a Gaussian approximation to the posterior (MacKay 1991, 1992b; Buntine and Weigend 1991) may break down as the number of hidden units becomes large. Markov chain Monte Carlo methods (Neal 1992b, 1993a, 1993b, and Chapter 3 of this book) produce the correct answer eventually, but may sometimes fail to reach the true posterior distribution in a reasonable length of time. In this chapter, I do not discuss such computational issues; my aim instead is to gain insight through theoretical analysis, done with varying degrees of rigour, and by sampling from the prior, which is much easier than sampling from the posterior.

For most of this chapter, I consider only multilayer perceptron networks that take I real-valued inputs, x_i, and produce O real-valued outputs given by functions $f_k(x)$, all of which are computed using a common layer of H hidden units, whose values $h_j(x)$, are produced with the tanh activation function. In detail:

$$f_k(x) = b_k + \sum_{j=1}^{H} v_{jk} h_j(x) \qquad (2.1)$$

$$h_j(x) = \tanh\left(a_j + \sum_{i=1}^{I} u_{ij} x_i\right) \qquad (2.2)$$

At times, I will consider networks in which the tanh activation function is replaced by a step function that takes the value -1 for negative arguments and $+1$ for positive arguments. (Learning for networks with step-function hidden units is computationally difficult, but these networks are sometimes simpler to analyse.) Networks with more than one hidden layer are discussed in Section 2.3.

When neural networks are used as regression and classification models, the outputs of the network are used to define the conditional distribution for the targets given the inputs. What is of interest then is the prior over these conditional distributions that results from the prior over output functions. For regression models, the relationship of the target distribution to the network outputs is generally simple — the outputs give the mean of a Gaussian distribution for the targets. For classification models such as the "softmax" model of Bridle (1989), the relationship is less straightforward. This matter is not examined in detail in this chapter; I look only at the properties of the prior over output functions, which provides the basis for understanding the prior over conditional distributions.

2.1 Priors converging to Gaussian processes

Most past work on Bayesian inference for neural networks (eg, MacKay 1992b) has used independent Gaussian distributions as the priors for network weights and biases. In this section, I investigate the properties of priors in which the hidden-to-output weights, v_{jk}, and output biases, b_k, have zero-mean Gaussian distributions with standard deviations of σ_v and σ_b. It will turn out that as the number of hidden units increases, the prior over functions implied by such priors converges to a Gaussian process. These priors can have smooth, Brownian, or fractional Brownian properties, as determined by the covariance function of the Gaussian process.

For the priors that I consider in detail, the input-to-hidden weights, u_{ij}, and the hidden unit biases, a_j, also have Gaussian distributions, with standard deviations σ_u and σ_a, though for the fractional Brownian priors, σ_u and σ_a are not fixed, but depend on the value of common parameters associated with each hidden unit.

2.1.1 Limits for Gaussian and other priors with finite variance

To determine what prior over functions is implied by a Gaussian prior for network parameters, let us look first at the prior distribution of the value of output unit k when the network inputs are set to some particular values, $x^{(1)}$ — that is we look at the prior distribution of $f_k(x^{(1)})$ that is implied by the prior distributions for the weights and biases.

From equation (2.1), we see that $f_k(x^{(1)})$ is the sum of a bias and the weighted contributions of the H hidden units. Under the prior, each term in this sum is independent, and the contributions of the hidden units all have identical distributions. The expected value of each hidden unit's contribution is zero: $E[v_{jk}h_j(x^{(1)})] = E[v_{jk}]E[h_j(x^{(1)})] = 0$, since v_{jk} is independent of a_j and the u_{ij} (which determine $h_j(x^{(1)})$), and $E[v_{jk}]$ is zero by hypothesis. The variance of the contribution of each hidden unit is finite: $E[(v_{jk}h_j(x^{(1)}))^2] = E[v_{jk}^2]E[h_j(x^{(1)})^2] = \sigma_v^2 E[h_j(x^{(1)})^2]$, which must be finite since $h_j(x^{(1)})$ is bounded. Defining $V(x^{(1)}) = E[h_j(x^{(1)})^2]$, which is the same for all j, we can conclude by the Central Limit Theorem that for large H the total contribution of the hidden units to the value of $f_k(x^{(1)})$ becomes approximately Gaussian, with variance $H\sigma_v^2 V(x^{(1)})$. The bias, b_k, is also Gaussian, with variance σ_b^2, so for large H the prior distribution of $f_k(x^{(1)})$ is also approximately Gaussian, with variance $\sigma_b^2 + H\sigma_v^2 V(x^{(1)})$.

Accordingly, to obtain a well-defined limit for the prior distribution of the value of the function at any particular point, we need only scale the prior variance of the hidden-to-output weights according to the number of hidden units, setting $\sigma_v = \omega_v H^{-1/2}$, for some fixed ω_v. The prior for $f_k(x^{(1)})$ then converges to a Gaussian of mean zero and variance $\sigma_b^2 + \omega_v^2 V(x^{(1)})$ as H goes to infinity.

Adopting this scaling for σ_v, we can investigate the prior joint distribution of the values of output k for several values of the inputs — that is, the joint distribution of $f_k(x^{(1)}), \ldots, f_k(x^{(n)})$, where $x^{(1)}, \ldots, x^{(n)}$ are the particular input values we choose to look at. An argument paralleling that above shows that as H goes to infinity this prior joint distribution converges to a multivariate Gaussian, with means of zero, and covariances

$$E[f_k(x^{(p)})f_k(x^{(q)})] = \sigma_b^2 + \sum_j \sigma_v^2 E[h_j(x^{(p)})h_j(x^{(q)})] \quad (2.3)$$

$$= \sigma_b^2 + \omega_v^2 C(x^{(p)}, x^{(q)}) \quad (2.4)$$

where $C(x^{(p)}, x^{(q)}) = E[h_j(x^{(p)})h_j(x^{(q)})]$, which is the same for all j. Distributions over functions of this sort, in which the joint distribution of the values of the function at any finite number of points is multivariate Gaussian, are known as *Gaussian processes*; they arise in many contexts,

2.1 Priors converging to Gaussian processes 33

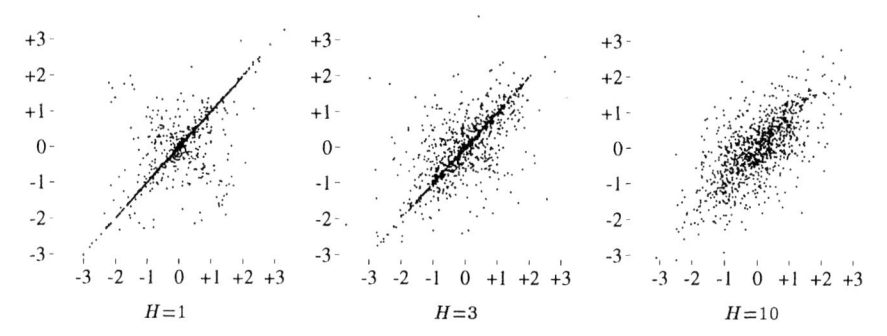

FIGURE 2.1. Convergence of network priors to a Gaussian process. Each of the plots is based on 1000 networks with one input unit, one output unit, and a single layer of 1, 3, or 10 tanh hidden units. The network weights were randomly drawn from prior distributions with $\sigma_u = 5$, $\sigma_a = 5$, $\sigma_b = 0.1$, and $\sigma_v = H^{-1/2}$, where H is the number of hidden units. Each network is represented by a point whose horizontal coordinate is the output of the network when the input is -0.2, and whose vertical coordinate is the output of the network when the input is $+0.4$.

including spatial statistics (Ripley 1981), splines (Wahba 1990), computer vision (Szeliski 1989), and computer graphics (Peitgen and Saupe 1988).

The prior covariances between the values of output k for different values of the inputs are in general not zero, which is what allows learning to occur. Given values for $f_k(x^{(1)}), \ldots, f_k(x^{(n-1)})$, we could explicitly find the infinite network's predictive distribution for the value of output k for case n by conditioning on these known values to produce a Gaussian distribution for $f_k(x^{(n)})$. This procedure may indeed be of practical interest, though it does require evaluation of $C(x^{(p)}, x^{(q)})$ for all $x^{(p)}$ in the training set and $x^{(q)}$ in the training and test sets, which would likely have to be done by numerical integration. In this book, predictive distributions for models based on finite networks will be found by other means (see Chapter 3), but insight into Bayesian learning for large networks with Gaussian priors can be gained by considering this picture of how a predictive distribution is formed by conditioning on the training data.

Figure 2.1 illustrates the convergence of network priors to a Gaussian process. The joint distribution of the network output for two particular input values is very non-Gaussian for a network with a single hidden unit ($H = 1$, on the left), but approaches a bivariate Gaussian distribution as the number of hidden units increases to $H = 3$ (middle) and $H = 10$ (on the right). Note that in the limiting prior distribution, the outputs of the network for these two inputs are correlated, so knowing the value of the output for one of these input values (a "training case") will help in predicting the output for the other input value (a "test case").

The joint distribution for the values of *all* the outputs of the network for some selection of values for inputs will also become a multivariate Gaussian in the limit as the number of hidden units goes to infinity. It is easy to see, however, that the covariance between $f_{k_1}(x^{(p)})$ and $f_{k_2}(x^{(q)})$ is zero whenever $k_1 \neq k_2$, since the weights into different output units are independent under the prior. Since zero covariance implies independence for Gaussian distributions, knowing the values of one output for various inputs does *not* tell us anything about the values of other outputs, at these or any other input points. When the number of hidden units is infinite, it makes no difference whether we train one network to produce two outputs, or instead use the same data to train two networks, each with one output. (I assume here that these outputs are not linked in some other fashion, such as by the assumption that their values are observed with a common, but unknown, level of noise.)

This independence of different outputs is perhaps surprising, since the outputs are computed using shared hidden units. However, with the Gaussian prior used here, the values of the hidden-to-output weights all go to zero as the number of hidden units goes to infinity. The output functions are built up from a large number of contributions from hidden units, with each contribution being of negligible significance by itself. Hidden units computing common features of the input that would be capable of linking the outputs are therefore not present. Dependencies between outputs could be introduced by making the weights to various outputs from one hidden unit be dependent, but if these weights have Gaussian priors, they can be dependent only if they are correlated. Accordingly, it is not possible to define a Gaussian-based prior expressing the idea that two outputs might show a large change in the same input region, the location of this region being unknown *a priori*, without also fixing whether the changes in the two outputs have the same or opposite sign.

The results in this section in fact hold more generally for any hidden unit activation function that is bounded, and for any prior on input-to-hidden weights and hidden unit biases (the u_{ij} and a_j) in which the weights and biases for different hidden units are independent and identically distributed. The results also apply when the prior for hidden-to-output weights is not Gaussian, as long as the prior has zero mean and finite variance.

2.1.2 *Priors that lead to smooth and Brownian functions*

I will start the detailed examination of Gaussian process priors by considering those that result when the input-to-hidden weights and hidden biases have Gaussian distributions. These turn out to give locally Brownian priors if step-function hidden units are used, and priors over smooth functions if tanh hidden units are used. For simplicity, I at first discuss only networks having a single input, but Section 2.1.5 will show that the results apply with little change to networks with any number of inputs.

2.1 Priors converging to Gaussian processes

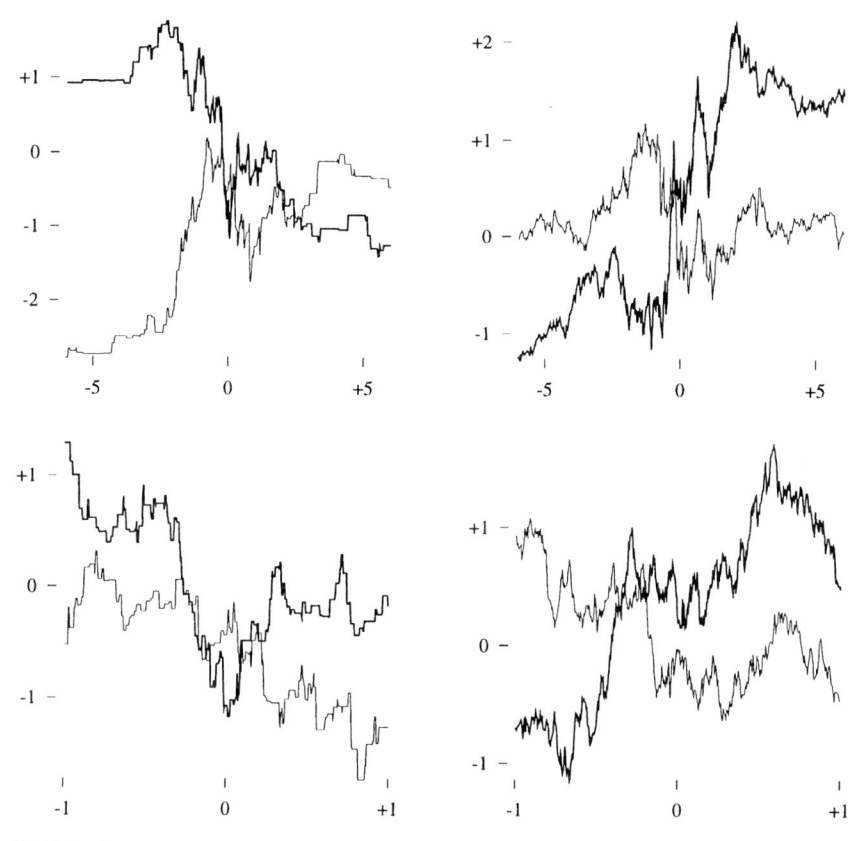

FIGURE 2.2. Functions drawn from Gaussian priors for networks with step function hidden units. The two functions shown on the left are from a network with 300 hidden units, the two on the right from a network with 10 000 hidden units. In both cases, $\sigma_a = \sigma_u = \sigma_b = \omega_v = 1$. The upper plots show the overall shape of each function; the lower plots show the central area in more detail.

To begin, consider a network with one input in which the hidden units compute a step function changing from -1 to $+1$ at zero. In this context, the values of the input weight, u_{1j}, and bias, a_j, for hidden unit j are significant only in that they determine the point in the input space where that hidden unit's step occurs, namely $-a_j/u_{1j}$. When the weight and bias have independent Gaussian prior distributions with standard deviations σ_u and σ_a, the prior distribution of this step-point is Cauchy, with a width parameter of σ_a/σ_u.

Figure 2.2 shows functions drawn from the prior distributions for two such networks, one network with 300 hidden units and one with 10 000 hidden units. Note that the general nature of the functions is the same for the two network sizes, but the functions from the larger network have more fine detail. This illustrates that the prior over functions is reaching a limiting distribution as H increases.

(In this and subsequent figures, the functions shown are not necessarily the first that were generated. Some selection was done in order to ensure that typical features are displayed, and to find pairs of functions that fit together nicely on a graph, without overlapping too much. In all cases, the functions shown were selected from a sample of no more than ten functions drawn from the prior.)

The variation in the functions shown in Figure 2.2 is concentrated in the region around $x = 0$, with a width of roughly σ_a/σ_u. Within this region, the function is locally Brownian in character, as a consequence of being built up from the many small, independent steps contributed by the hidden units. Far from $x = 0$, the functions become almost constant, since few hidden units have their steps that far out. For the remainder of this chapter, I will confine my attention to the properties of functions in their central regions, where all points have approximately equal potential for being influenced by the hidden units.

When tanh hidden units are used instead of step-function units, the functions generated are smooth. This can be seen by noting that all the derivatives (to any order) of the value of a tanh hidden unit with respect to the inputs are polynomials in the hidden unit value and the input-to-hidden weights. These derivatives therefore have finite expectation and finite variance, since the hidden unit values are bounded, and the weights are from Gaussian distributions, for which moments of all orders exist. At scales greater than about $1/\sigma_u$, however, the functions exhibit the same Brownian character that was seen with step-function hidden units.

The size of the central region where the properties of these functions are approximately uniform is roughly $(\sigma_a + 1)/\sigma_u$. To see this, note that when the input weight is u, the distribution of the point where the hidden unit value crosses zero is Gaussian with standard deviation $\sigma_a/|u|$. The influence of a hidden unit with this input weight extends a distance of about $1/|u|$, however, so points within about $(\sigma_a + 1)/|u|$ of the origin are potentially influenced by hidden units with input weights of this size. Since the probability of obtaining a weight of size $|u|$ declines exponentially beyond $|u| = \sigma_u$, the functions will have similar properties at all points within a distance of about $(\sigma_a + 1)/\sigma_u$ of the origin.

Functions drawn from priors for networks with tanh hidden units are shown in Figure 2.3.

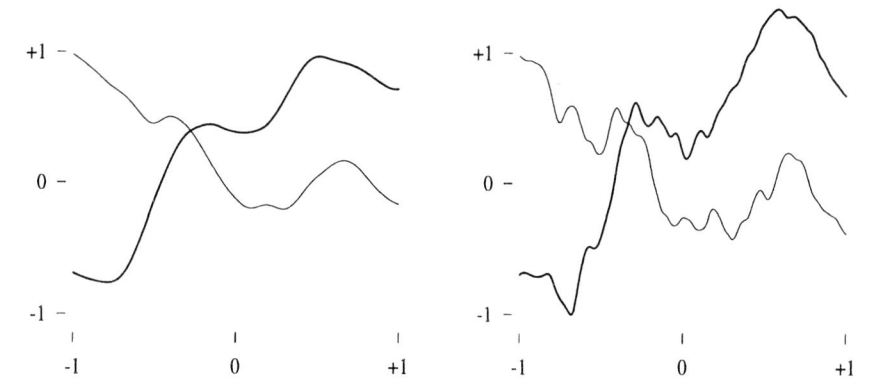

FIGURE 2.3. Functions drawn from Gaussian priors for a network with 10 000 tanh hidden units. Two functions drawn from a prior with $\sigma_u = 5$ are shown on the left, two from a prior with $\sigma_u = 20$ on the right. In both cases, $\sigma_a/\sigma_u = 1$ and $\sigma_b = \omega_v = 1$. The functions with different σ_u were generated using the same random number seed, the same as that used to generate the functions in the lower-right of Figure 2.2. This allows a direct evaluation of the effect of changing σ_u. (Note that use of step function hidden units is equivalent to letting σ_u go to infinity, while keeping σ_a/σ_u fixed.)

2.1.3 Covariance functions of Gaussian priors

A Gaussian process can be completely characterized by the mean values of the function at each point, always zero for the network priors discussed here, along with the covariance of the function value at any two points, given by equation (2.4). The difference between priors that lead to locally smooth functions and those that lead to locally Brownian functions is reflected in the local behaviour of their covariance functions. From equation (2.4), we see that this is directly related to the covariance of the values of a hidden unit at nearby input points, $C(x^{(p)}, x^{(q)})$, which can be written as

$$C(x^{(p)}, x^{(q)}) = \tfrac{1}{2}\left(V(x^{(p)}) + V(x^{(q)}) - E\big[(h(x^{(p)}) - h(x^{(q)}))^2\big]\right) \quad (2.5)$$
$$= V - \tfrac{1}{2} D(x^{(p)}, x^{(q)}) \quad (2.6)$$

where $V(x^{(p)}) \approx V \approx V(x^{(q)})$, for nearby $x^{(p)}$ and $x^{(q)}$, and $D(x^{(p)}, x^{(q)})$ is the expected squared difference between the values of a hidden unit at $x^{(p)}$ and $x^{(q)}$.

For step-function hidden units, $\big(h(x^{(p)}) - h(x^{(q)})\big)^2$ will be either 0 or 4, depending on whether the values of the hidden unit's bias and incoming weight result in the step being located between $x^{(p)}$ and $x^{(q)}$. Since the location of this step will be approximately uniform in the local vicinity, the probability of the step occurring between $x^{(p)}$ and $x^{(q)}$ will rise

38   Chapter 2. Priors for Infinite Networks

proportionally with the separation of the points, giving

$$D(x^{(p)}, x^{(q)}) \sim |x^{(p)} - x^{(q)}| \tag{2.7}$$

where $\sim$ indicates proportionality for nearby points. This behaviour is characteristic of Brownian motion.

For networks with tanh hidden units, with Gaussian priors for the bias and incoming weight, we have seen that the functions are smooth. Accordingly, for nearby $x^{(p)}$ and $x^{(q)}$ we will have

$$D(x^{(p)}, x^{(q)}) \sim |x^{(p)} - x^{(q)}|^2 \tag{2.8}$$

We can get a rough idea of the behaviour of $D(x^{(p)}, x^{(q)})$ for all points within the central region as follows. First, fix the input-to-hidden weight, u, and consider the expectation of $\bigl(h(x-s/2) - h(x+s/2)\bigr)^2$ with respect to the prior distribution of the bias, a, which is Gaussian with standard deviation σ_a. With u fixed, the point where the hidden unit's total input crosses zero will have a prior distribution that is Gaussian with standard deviation $\sigma_a/|u|$, giving a probability density for the zero crossing to occur at any point in the central region of around $|u|/\sigma_a$. We can now distinguish two cases. When $|u| \gtrsim 1/s$, the transition region over which the hidden unit's output changes from -1 to $+1$, whose size is about $1/|u|$, will be small compared to s, and we can consider that $\bigl(h(x-s/2) - h(x+s/2)\bigr)^2$ will either be 0 or 4, depending on whether the total input to the hidden unit crosses zero between $x-s/2$ and $x+s/2$, which occurs with probability around $(|u|/\sigma_a)s$. When $|u| \lesssim 1/s$, $\bigl(h(x-s/2) - h(x+s/2)\bigr)^2$ will be about $(|u|s)^2$ if the interval $[x-s/2, x+s/2]$ is within the transition region, while otherwise it will be nearly zero. The probability of $[x-s/2, x+s/2]$ lying in the transition region will be about $(|u|/\sigma_a)(1/|u|) = 1/\sigma_a$. Putting all this together, we get

$$E_a\bigl[\bigl(h(x-s/2) - h(x+s/2)\bigr)^2\bigr] \approx \begin{cases} c_1(|u|/\sigma_a)s & \text{if } |u| \gtrsim 1/s \\ c_2(|u|^2/\sigma_a)s^2 & \text{if } |u| \lesssim 1/s \end{cases} \tag{2.9}$$

where $c_1, c_2, \ldots$ are constants of order one. Taking the expectation with respect to a symmetrical prior for u, with density $p(u)$, we get

$$E_{a,u}\bigl[\bigl(h(x-s/2) - h(x+s/2)\bigr)^2\bigr]$$

$$\approx 2\frac{c_1 s}{\sigma_a}\int_{1/s}^{\infty} u\, p(u)\, du + 2\frac{c_2 s^2}{\sigma_a}\int_0^{1/s} u^2\, p(u)\, du \tag{2.10}$$

Finally, if we crudely approximate the Gaussian prior for u by a uniform distribution over $[-\sigma_u, +\sigma_u]$, with density $p(u) = 1/2\sigma_u$, we get

$$D(x-s/2, x+s/2) = E_{a,u}\bigl[\bigl(h(x-s/2) - h(x+s/2)\bigr)^2\bigr]$$

$$\approx \frac{1}{\sigma_a} \begin{cases} c_3 \, \sigma_u^2 s^2 & \text{if } s \lesssim 1/\sigma_u \\ c_4 \, \sigma_u s + c_5/\sigma_u s & \text{if } s \gtrsim 1/\sigma_u \end{cases} \quad (2.11)$$

Thus these functions are smooth on a small scale, but when viewed on scales significantly larger than $1/\sigma_u$, they have a Brownian nature characterized by $D(x-s/2, x+s/2)$ being proportional to s.

2.1.4 Fractional Brownian priors

It is natural to wonder whether a prior on the weights and biases going into hidden units can be found for which the resulting prior over functions has *fractional Brownian* properties (Falconer 1990, Section 16.2), for which

$$D(x^{(p)}, x^{(q)}) \sim |x^{(p)} - x^{(q)}|^\eta \quad (2.12)$$

As above, values of $\eta = 2$ and $\eta = 1$ correspond to smooth and Brownian functions, respectively. Functions with intermediate properties are obtained when $1 < \eta < 2$; functions "rougher" than Brownian functions are obtained when $0 < \eta < 1$.

One way to achieve these effects would be to change the hidden unit activation function from $\tanh(z)$ to $\text{sign}(z)|z|^{(\eta-1)/2}$ (Peitgen and Saupe 1988, Sections 1.4.1 and 1.6.11). However, the unbounded derivatives of this activation function would pose problems for gradient-based learning methods. I will describe a method of obtaining fractional Brownian functions with $1 < \eta < 2$ from networks with tanh hidden units by altering the priors for the hidden unit biases and input weights.

To construct this fractional Brownian prior, we associate with hidden unit j an "adjustment" value, A_j, that controls the magnitude of that hidden unit's incoming weights and bias. Given A_j, we let the incoming weights, u_{ij}, have independent Gaussian distributions with standard deviation $\sigma_u = A_j \omega_u$, and we let the bias, a_j, have a Gaussian distribution with standard deviation $\sigma_a = A_j \omega_a$. We give the A_j themselves independent prior distributions with probability density $p(A) \propto A^{-\eta} \exp\left(-(\eta-1)/2A^2\right)$, where $\eta > 1$, which corresponds to a Gamma distribution for $1/A_j^2$. Note that if we integrate over A_j to obtain a direct prior for the weights and biases, we find that the weights and biases are no longer independent, and no longer have Gaussian distributions.

To picture why this setup should result in a fractional Brownian prior for the functions computed by the network, consider that when A_j is large, $h_j(x)$ is likely to be almost a step function, since σ_u will be large. (A_j does not affect the distribution of the point where the step occurs, however, since this depends only on σ_a/σ_u.) Such near-step-functions produced by hidden units with A_j greater than some limit will contribute in a Brownian fashion to $D(x^{(p)}, x^{(q)})$, with the contribution rising in direct proportion

40   Chapter 2. Priors for Infinite Networks

to the separation of $x^{(p)}$ and $x^{(q)}$. However, as this separation increases, the value of A_j that is sufficient for the hidden unit to behave as a step function in this context falls, and the number of hidden units that effectively behave as step functions rises. The contribution of such hidden units to $D(x^{(p)}, x^{(q)})$ will therefore increase faster than for a Brownian function. The other hidden units with small A_j will also contribute to $D(x^{(p)}, x^{(q)})$, quadratically with separation, but for nearby points their contribution will be dominated by that of the units with large A_j, if that contribution is sub-quadratic.

We can see this in somewhat more detail by substituting $\sigma_u = A_j \omega_u$ and $\sigma_a = A_j \omega_a$ in equation (2.11), obtaining

$$E_{a,u}\left[(h(x-s/2) - h(x+s/2))^2\right]$$

$$\approx \frac{1}{\omega_a} \begin{cases} c_3 A_j \omega_u^2 s^2 & \text{if } A_j \lesssim 1/s\omega_u \\ c_4 \omega_u s + c_5/A_j^2 \omega_u s & \text{if } A_j \gtrsim 1/s\omega_u \end{cases} \quad (2.13)$$

Integrating with respect to the prior for A_j, we get

$$D(x-s/2, x+s/2) \approx \frac{c_3 \omega_u^2 s^2}{\omega_a} \int_0^{1/\omega_u s} A\, p(A)\, dA + \frac{c_4 \omega_u s}{\omega_a} \int_{1/\omega_u s}^{\infty} p(A)\, dA$$

$$+ \frac{c_5}{\omega_a \omega_u s} \int_{1/\omega_u s}^{\infty} A^{-2} p(A)\, dA \quad (2.14)$$

The mode of $p(A)$ is at $((\eta-1)/\eta)^{1/2}$. Before this point $p(A)$ drops rapidly, and can be approximated as being zero; after this point, it drops as $A^{-\eta}$. The integrals above can thus be approximated as follows, for $\eta \neq 2$:

$$D(x-s/2, x+s/2)$$

$$\approx \frac{1}{\omega_a} \begin{cases} c_6 \omega_u^\eta s^\eta + c_7 \omega_u^2 s^2 & \text{if } s \lesssim (\eta/(\eta-1))^{1/2}/\omega_u \\ c_8 \omega_u s + c_9/\omega_u s & \text{if } s \gtrsim (\eta/(\eta-1))^{1/2}/\omega_u \end{cases} \quad (2.15)$$

When $1 < \eta < 2$, the s^η term will dominate for small s, and the function will have fractional Brownian properties; when $\eta > 2$, the s^2 term will dominate, producing a smooth function; $\eta = 2$ is a special case, for which $D(x-s/2, x+s/2) \sim s^2 \log(1/s)$.

Fractional Brownian functions drawn from these priors are shown in Figure 2.4. Figure 2.5 shows the behaviour of $D(x-s/2, x+s/2)$ for the same priors, as well as for the priors used in Figures 2.2 and 2.3.

2.1.5 Networks with more than one input

The priors discussed here have analogous properties when used for networks with several inputs. In particular, the value of the network function along

2.1 Priors converging to Gaussian processes   41

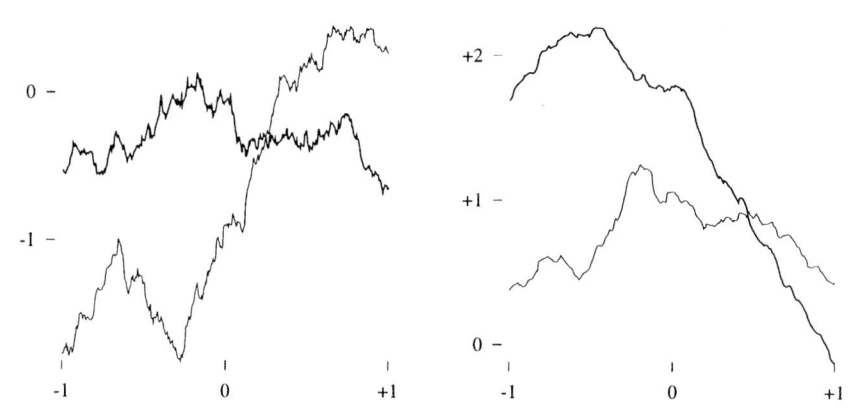

FIGURE 2.4. Functions drawn from fractional Brownian priors for a network with 10 000 tanh hidden units. Two functions drawn from a prior with $\eta = 1.3$ are shown on the left, two from a prior with $\eta = 1.7$ on the right. In both cases, $\omega_a = \omega_u = \sigma_b = \omega_v = 1$.

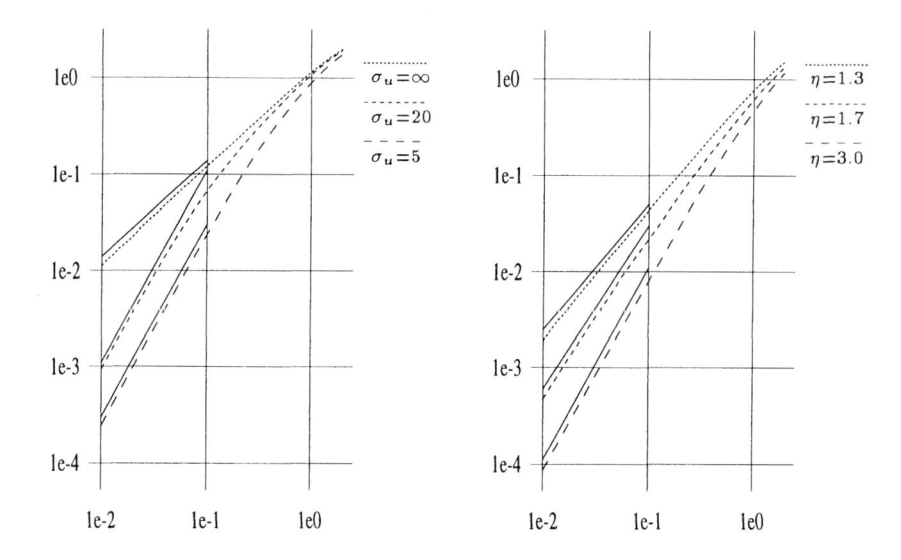

FIGURE 2.5. Behaviour of $D(x-s/2, x+s/2)$ as s varies for Brownian, smooth, and fractional Brownian functions. The plots on the left are for the Brownian prior used in Figure 2.2, and the smooth priors used in Figure 2.3; those on the right are for the fractional Brownian priors used in Figure 2.4, as well as for a similar prior on the A_j with $\eta = 3$, which leads to a smooth function. All values are for $x = 0.2$. They were computed by Monte Carlo integration using a sample of 100 000 values drawn from the prior for the bias and weight into a hidden unit; the values are hence subject to a small amount of noise. Note that both scales are logarithmic, so that a function proportional to s^η should appear as a straight line of slope η. Straight lines of the expected slopes are shown beside the curves to demonstrate their convergence for small s.

42   Chapter 2. Priors for Infinite Networks

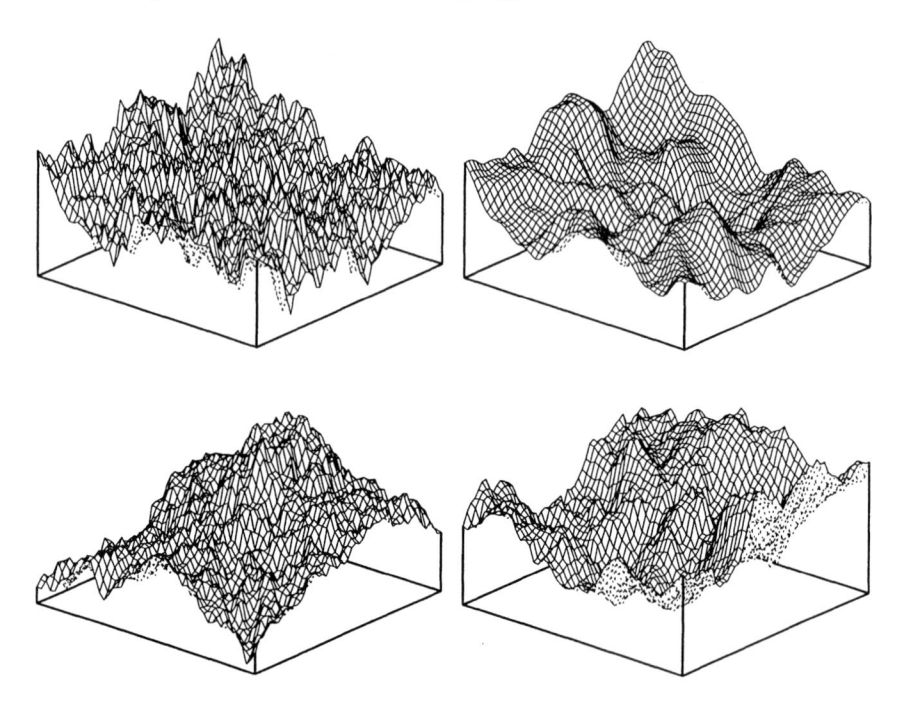

FIGURE 2.6. Functions of two inputs drawn from Gaussian priors. The function in the upper left is from a network with 10 000 step-function hidden units, that in the upper right from the corresponding network with tanh hidden units, using the same random number seed. In both cases, $\sigma_a = \sigma_u = 10$. The two lower functions are from networks with tanh hidden units, using fractional Brownian priors. The function in the lower left has $\eta = 1.3$, that in the lower right $\eta = 1.7$. In both cases, $\omega_a = \omega_u = 1$. The plots show the input region from -1 to $+1$.

any line in input space has the same properties as those described above for a network with a single input. Since all the priors discussed are invariant with respect to rotations of the input space, we may confine our attention to lines obtained by varying only one of the inputs, say the first. Rewriting equation (2.2) as

$$h_j(x) = \tanh\left(u_{1j}x_1 + a_j + \sum_{i=2}^{I} u_{ij}x_i\right) \qquad (2.16)$$

we see that when $x_2, \ldots, x_I$ are fixed, they act simply to increase the variance of the effective bias. This merely spreads the variation in the function over a larger range of values for x_1.

Figure 2.6 shows functions of two inputs drawn from Brownian, smooth, and fractional Brownian priors.

2.2 Priors converging to non-Gaussian stable processes

Although we have seen that a variety of interesting priors over functions can be produced using Gaussian priors for hidden-to-output weights and output biases, these priors are in some respects disappointing.

One reason for this is that it may be possible to implement Bayesian inference for these priors, or for other Gaussian process priors with similar properties, using standard methods based directly on the covariance function, without any need for an actual network. We may thus need to look at different priors if Bayesian neural networks are to significantly extend the range of models available. (On the other hand, it is possible that the particular covariance function created by the network might be of special interest, or that control of the covariance function via hyperparameters might most conveniently be done in a network formulation.)

Furthermore, as mentioned earlier, with Gaussian priors the contributions of individual hidden units are all negligible, and consequently, these units do not represent "hidden features" that capture important aspects of the data. If we wish the network to do this, we need instead a prior with the property that even in the limit of infinitely many hidden units, there are some individual units that have non-negligible output weights. Such non-Gaussian priors can indeed be constructed, using prior distributions for the weights from hidden to output units that do not have finite variance.

2.2.1 Limits for priors with infinite variance

The theory of *stable distributions* (Feller 1966, Section VI.1, Samorodnitsky and Taqqu 1994) provides the basis for analysing the convergence of priors in which hidden-to-output weights have infinite variance. If random variables $Z_1, \ldots, Z_n$ are independent, and all the Z_i have the same symmetric stable distribution of index α, then $(Z_1 + \cdots + Z_n)/n^{1/\alpha}$ has the same distribution as the Z_i. Such symmetric stable distributions exist for $0 < \alpha \leq 2$, and for each index they form a single family, varying only in width. The symmetric stable distributions of index $\alpha = 2$ are the Gaussians of varying standard deviations; those of index $\alpha = 1$ are the Cauchy distributions of varying widths; the densities for the symmetric stable distributions with most other indexes have no convenient forms.

If independent variables $Z_1, \ldots, Z_n$ each have the same distribution, one that is in the *normal domain of attraction* of the family of symmetric stable distributions of index α, then the distribution of $(Z_1 + \cdots + Z_n)/n^{1/\alpha}$ approaches such a stable distribution as n goes to infinity. All distributions with finite variance are in the normal domain of attraction of the Gaussian. Distributions with tails that (roughly speaking) have densities that decline as $z^{-(\alpha+1)}$, with $0 < \alpha < 2$ are in the normal domain of attraction of the symmetric stable distributions of index α (Feller, 1966, Sections IX.8 and XVII.5).

We can define a prior on network weights in such a fashion that the resulting prior on the value of a network output for a particular input converges to a non-Gaussian symmetric stable distribution as the number of hidden units, H, goes to infinity. This is done by using independent, identical priors for the hidden-to-output weights, v_{jk}, with a density whose tails go as $v_{jk}^{-(\alpha+1)}$, with $\alpha < 2$. For all the examples in this book, I use a t-distribution with density proportional to $(1 + v_{jk}^2/\alpha\sigma_v^2)^{-(\alpha+1)/2}$. The prior distribution of the contribution of a hidden unit to the output will have similar tail behaviour, since the hidden unit values are bounded. Accordingly, if we scale the width parameter of the prior for hidden-to-output weights as $\sigma_v = \omega_v H^{-1/\alpha}$, the prior for the total contribution of all hidden units to the output value for a particular input will converge to a symmetric stable distribution of index α. If the prior for the output bias is a stable distribution of this same index, the value of the output unit for that input, which is the sum of the bias and the hidden unit contributions, will have a prior distribution in this same stable family. (In practice, it may not be convenient for the bias to have such a stable distribution as its prior, but using a different prior for the bias will have only a minor effect.)

To rigorously show that these priors converge, we would need to show not only that the prior distribution for the value of the function at any single point converges (as shown above), but that the joint distribution of the value of the function at any number of points converges as well — i.e. that the dependencies between points converge. I do not attempt this here, but the plots below (e.g. Figure 2.7) lend empirical support to this proposition.

To gain insight into the nature of priors based on non-Gaussian stable distributions, we can look at the expected number of hidden-to-output weights lying in some small interval, $[w, w + \epsilon]$, in the limit as H goes to infinity. For a given H, the number of weights in this interval using the prior that is scaled down by $H^{-1/\alpha}$ will be the same as the number that would be in the interval $[wH^{1/\alpha}, wH^{1/\alpha} + \epsilon H^{1/\alpha}]$ if the unscaled prior were used. As H increases, this interval moves further and further into the tail of the unscaled prior distribution, where, by construction, the density goes down as $v^{-(\alpha+1)}$. The probability that a particular weight will lie in this small interval is thus proportional to $\epsilon H^{1/\alpha}(wH^{1/\alpha})^{-(\alpha+1)} = \epsilon w^{-(\alpha+1)} H^{-1}$. The expected total number of weights from all H hidden units that lie in the interval $[w, w + \epsilon]$ is therefore proportional to $\epsilon w^{-(\alpha+1)}$, in the limit as H goes to infinity.

2.2 Priors converging to non-Gaussian stable processes

Thus, whereas for Gaussian priors, all the hidden-to-output weights go to zero as H goes to infinity, for priors based on symmetric stable distributions of index $\alpha < 2$, some of the hidden units in an infinite network have output weights of significant size, allowing them to represent "hidden features". As an aside, the fact that the number of weights of each size has non-zero expectation means that the prior can be given an alternative formulation in terms of a Poisson process for hidden-to-output weights. (Note that though such a process could be defined for any α, it gives rise to a well-defined prior over functions only if $0 < \alpha < 2$.)

The above priors based on non-Gaussian stable distributions lead to prior distributions over functions in which the functions computed by different output units are independent, in the limit as H goes to infinity, just as was the case for Gaussian priors. This comes about because the weights to the various output units from a single hidden unit are independent. As H goes to infinity, the fraction of weights that are of significant size goes to zero, even while the actual number of such weights remains non-zero. There is thus a vanishingly small chance that a single hidden unit will have a significant effect on two different outputs, which is what would be needed to make the two outputs dependent.

However, with non-Gaussian priors, we can introduce dependence between outputs without also introducing correlation. One way to do this is use t-distributions that are expressed as mixtures of Gaussian distributions of varying scale. With each hidden unit, j, we associate an output weight variance hyperparameter, $\sigma_{v,j}^2$. As a prior, we give $1/\sigma_{v,j}^2$ a Gamma distribution with shape parameter $\alpha/2$ and mean σ_v. Given a value for this common hyperparameter, the weights out of a hidden unit, v_{jk}, have independent Gaussian distributions of variance $\sigma_{v,j}^2$. By integrating over the hyperparameter, one can see that each hidden-to-output weight has a t-distribution with index α, as was the case above. Now, however, the weights out of a single hidden unit are dependent — they are all likely to have similar magnitudes, since they depend on the common value of σ_v. This prior thus allows single hidden units to compute common features that affect many outputs, without fixing whether these effects are in the same or different directions.

2.2.2 Properties of non-Gaussian stable priors

In contrast to the situation for Gaussian process priors, whose properties are captured by their covariance functions, I know of no simple way to characterize the distributions over functions produced by the priors based on non-Gaussian stable distributions. I will therefore confine myself in this section to illustrating the nature of these priors by displaying functions sampled from them.

46   Chapter 2. Priors for Infinite Networks

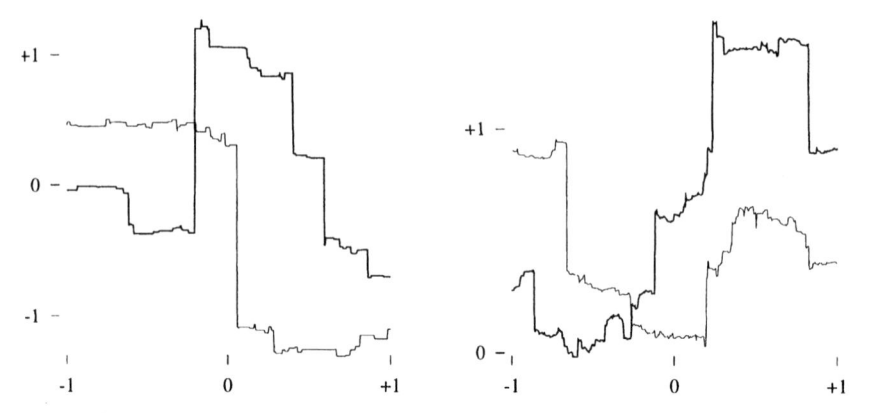

FIGURE 2.7. Functions drawn from Cauchy priors for networks with step-function hidden units. Functions shown on the left are from a network with 150 hidden units, those on the right from a network with 10 000 hidden units. In both cases, $\sigma_a = \sigma_u = \sigma_b = \omega_v = 1$.

As before, we can begin by considering a network with a single real input and a single real output, with step-function hidden units. Figure 2.7 shows two functions drawn from priors for such networks in which the weights and biases into the hidden units have independent Gaussian distributions and the weights and bias for the output have Cauchy distributions (the stable distribution with $\alpha = 1$). Networks with 150 hidden units and with 10 000 hidden units are shown, for which the width parameter of the Cauchy distribution was scaled as $\sigma_v = \omega_v H^{-1}$. As is the case for the Gaussian priors illustrated in Figure 2.2, the general nature of the functions is the same for the small networks and the large networks, with the latter simply having more fine detail. The functions are clearly very different from those drawn from the Gaussian prior that are shown in Figure 2.2. The functions from the Cauchy prior have large jumps due to single hidden units that have output weights of significant size.

When the prior on hidden-to-output weights has a form that converges to a stable distribution with $0 < \alpha < 1$, the dominance of small numbers of hidden units becomes even more pronounced than for the Cauchy prior. For stable priors with $1 < \alpha < 2$, effects intermediate between the Cauchy and the Gaussian priors are obtained. These priors may of course be used in conjunction with tanh hidden units. Figure 2.8 illustrates some of these possibilities for functions of two inputs.

An infinite network whose prior is based on a stable distribution with a small α can be used to express whatever valid intuitions we may sometimes have that might otherwise lead us to use a network with a small number of hidden units. With a small α, the contributions of a small subset of the hidden units will dominate, which will be good if we in fact have reason to believe that the true function is close to one that can be represented by a

2.2 Priors converging to non-Gaussian stable processes   47

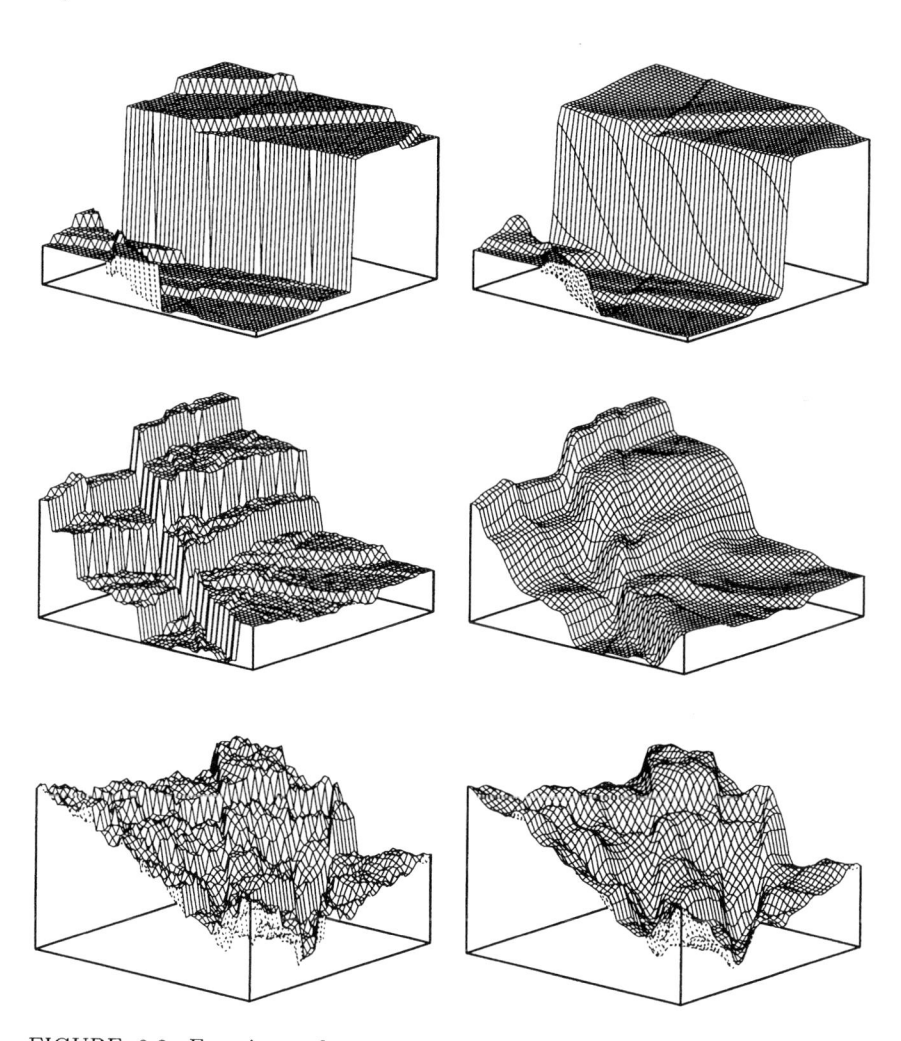

FIGURE 2.8. Functions of two inputs drawn from priors that converge to non-Gaussian stable distributions. Functions on the left are from networks with step-function hidden units; those on the right are the corresponding functions from networks with tanh hidden units, with $\sigma_u = 20$. For the functions at the top, the prior on hidden-to-output weights was a t-distribution with $\alpha = 0.5$; in the middle, the prior was Cauchy (a t-distribution with $\alpha = 1$); on the bottom the prior was a t-distribution with $\alpha = 1.5$. All the networks had 1000 hidden units. In all cases, priors with $\sigma_a/\sigma_u = 1$ were used; the plots extend from -1 to $+1$ for both inputs, within the corresponding central region.

48   Chapter 2. Priors for Infinite Networks

small network. The remaining hidden units will still be present, however, and able to make any small corrections that are needed to represent the function exactly.

2.3   Priors for nets with more than one hidden layer

In this section, I take a preliminary look at priors for multilayer perceptron networks with more than one layer of hidden units, starting with networks in which the outputs are connected only to the last hidden layer, each hidden layer after the first has incoming connections only from the preceding hidden layer, and the first hidden layer has incoming connections only from the inputs.

Consider such a network with several layers of step-function hidden units, with all the weights and biases having Gaussian prior distributions. Assume that the standard deviation of the weights on the connections out of a hidden layer with H units is scaled down by $H^{-1/2}$, as before. We are again interested in the limiting distribution over functions as the number of hidden units in each layer goes to infinity.

Figure 2.9 shows functions of one input drawn from this prior for networks with one, two, and three hidden layers. The function value is shown by a dot at each of 500 grid points in the central region of the input space. (This presentation shows the differences better than a line plot does.) With one hidden layer, the function is Brownian, as was already seen in Figure 2.2. With two hidden layers, the covariance between nearby points falls off much more rapidly with their separation, and with three hidden layers, this appears to be even more pronounced.

This is confirmed by numerical investigation, which shows that the networks with two and three hidden layers satisfy equation (2.12) with $\eta \approx 1/2$ and $\eta \approx 1/4$, respectively. For networks where only the first hidden layer is connected to the inputs, it should be true in general that adding an additional hidden layer with step-function units after what was previously the last hidden layer results in a reduction of η by a factor of two. To see this, note first that the total input to one of the hidden units in this new layer will have the same distribution as the output of the old network. For a unit in the new hidden layer, $\left(h(x^{(p)}) - h(x^{(q)})\right)^2$ will be 0 or 4 depending on whether the unit's total input changes sign between $x^{(p)}$ and $x^{(q)}$. The probability of this occurring will be directly proportional to the difference in value between the total input to the unit at $x^{(p)}$ and the total input at $x^{(q)}$. By hypothesis, this difference is Gaussian with a variance proportional to $|x^{(p)} - x^{(q)}|^\eta$, giving an expected absolute magnitude for the difference that is proportional to $|x^{(p)} - x^{(q)}|^{\eta/2}$. From this it

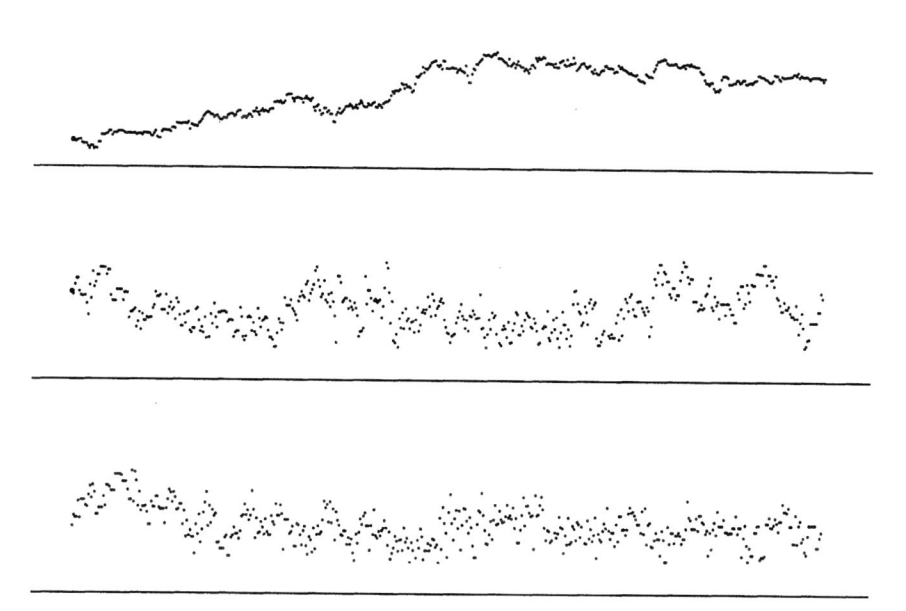

2.3 Priors for nets with more than one hidden layer 49

FIGURE 2.9. Functions computed by networks with one (top), two (middle), and three (bottom) layers of step-function hidden units, with Gaussian priors. All networks had 2000 units in each hidden layer. The value of each function is shown at 500 grid points along the horizontal axis.

follows that $D(x^{(p)}, x^{(q)}) = E\big[\big(h(x^{(p)}) - h(x^{(q)})\big)^2\big]$ is also proportional to $|x^{(p)} - x^{(q)}|^{\eta/2}$.

Though it is interesting that fractional Brownian priors with $\eta < 1$ can be obtained in this manner, I suspect that such priors will have few applications. For small values of η, the covariances between the function values at different points drop off rapidly with distance, introducing unavoidable uncertainty in predictions for test points that are even slightly different from training points. This situation is difficult to distinguish from that where the observed function values are subject to independent Gaussian noise, unless the training set contains multiple observations for exactly the same input values. Modeling independent noise is much easier than modeling fractional Brownian functions, and hence is to be preferred on pragmatic grounds when both models would give similar results.

More interesting effects can be obtained using a combination of Gaussian and non-Gaussian priors in a network with two hidden layers of the following structure. The first hidden layer contains H_1 tanh or step-function units, with priors for the biases and the weights on the input connections that are Gaussian, or of the fractional Brownian type described in Section 2.1.4. The second hidden layer contains H_2 tanh or step-function units, with Gaussian priors for the biases and for the weights on the connections

50   Chapter 2. Priors for Infinite Networks

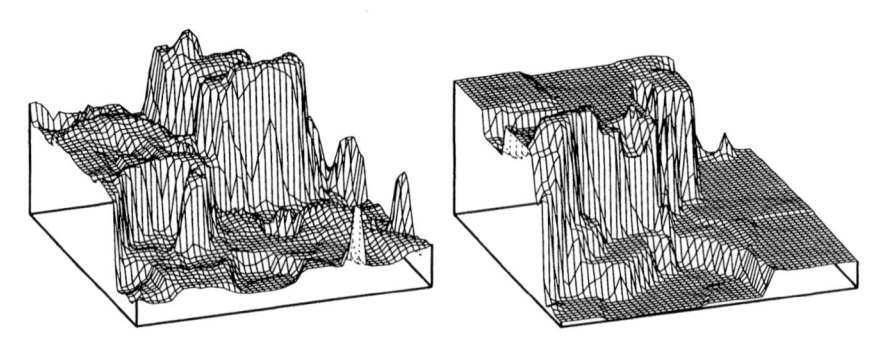

FIGURE 2.10. Two functions drawn from a combined Gaussian and non-Gaussian prior for a network with two layers of tanh hidden units. The first hidden layer contained $H_1 = 500$ units; the second contained $H_2 = 300$ units. The priors for weights and biases into the first hidden layer were Gaussian with standard deviation 10. The priors for weights and biases into the second hidden layer were also Gaussian, with the biases having standard deviation 20 and the weights from the first hidden layer having standard deviation $20H_1^{-1/2}$. The weights from the second hidden layer to the output were drawn from a t-distribution with $\alpha = 0.6$ and a width parameter of $H_2^{-1/0.6}$, which converges to the corresponding stable distribution. The central regions of the functions are shown, where the inputs vary from -1 to $+1$.

from the first hidden layer (with the standard deviation for these weights scaled as $H_1^{-1/2}$). There are no direct connections from the inputs to the second hidden layer. Finally, the outputs are connected only to the last hidden layer, with a prior for the hidden-to-output weights that converges to a non-Gaussian stable distribution of index α (for which the width of the prior will scale as $H_2^{-1/\alpha}$).

With this setup, the function giving the total input into a unit in the second hidden layer has the same prior distribution as the output function for a network of one hidden layer with Gaussian priors, which may, for example, have the forms seen in Figures 2.2, 2.3, 2.4, or 2.6. The step-function or tanh hidden units will convert such a function into one bounded between -1 and $+1$. Such a hidden unit may be seen as a "feature detector" that indicates whether the network inputs lie in one of the regions where the hidden unit's total input is significantly greater than zero. The use of non-Gaussian priors for the weights from these hidden units to the outputs allows individual features to have a significant effect on the output.

Functions drawn from such a prior are illustrated in Figure 2.10. Such functions have low probability under the priors for networks with one hidden layer that have been discussed, suggesting that two-layer networks will be advantageous in some applications.

Finally, we can consider the limiting behaviour of the prior over functions as the number of hidden layers increases. If the priors on hidden-to-hidden weights, hidden unit biases, and input-to-hidden weights (if present) are the same for all hidden layers, the prior over the functions computed by the units in the hidden layers of such a network will have the form of a homogeneous Markov chain — that is, under the prior, the distribution of functions computed by hidden units in layer $\ell + 1$ is influenced by the functions computed by earlier layers only through the functions computed by layer ℓ, and furthermore, the conditional distribution of functions computed by layer $\ell + 1$ given those computed by layer ℓ is the same for all ℓ. We can now ask whether this Markov chain converges to some invariant distribution as the number of layers goes to infinity, given the starting point established by the prior on weights into the first hidden layer. If the chain does converge, then the prior over functions computed by the output units should also converge, since the outputs are computed solely from the hidden units in the last layer.

This question of convergence appears difficult to answer. Indeed, when each hidden layer contains an infinite number of hidden units, it is not even obvious how convergence should be defined. Nevertheless, from the discussion above, it is clear that a Gaussian-based prior for a network with many layers of step-function hidden units, with no direct connections from inputs to hidden layers after the first, either does not converge as the number of layers goes to infinity, or if it can be regarded as converging, it is to an uninteresting distribution concentrated on completely unlearnable functions. However, if direct connections from the inputs to all the hidden layers are included, it appears that convergence to a sensible distribution may occur, and of course there are also many possibilities involving non-Gaussian stable priors and hidden units that compute a smooth function such as tanh rather than a step function.

Finding a prior with sensible properties for a network with an infinite number of hidden layers, each with an infinite number of units, would perhaps be the ultimate demonstration that Bayesian inference does not require limiting the complexity of the model. Whether such a result would be of any practical significance would of course depend on whether such networks have any significant advantage over networks with one or two layers, and on whether a prior close to the limit is obtained with a manageable number of layers (say less than ten) and a manageable number of hidden units per layer (at most in the hundreds).

2.4 Hierarchical models

Often, our prior knowledge will be too unspecific to fix values for σ_b, ω_v, σ_a (or ω_a), and σ_u (or ω_u), even if we have complete insight into their effects on the prior. We may then wish to treat these values as unknown hyperparameters, giving them higher-level prior distributions that are rather broad. Insight into the nature of the prior distributions produced for given values of the hyperparameters is still useful even when we plan to use such a hierarchical model, rather than fixing the hyperparameters to particular values, since this insight allows us to better understand the nature of the model, and to judge whether the range of possibilities it offers is adequate for our problem.

One benefit of a hierarchical model is that the degree of "regularization" that is appropriate for the task can be determined automatically from the data (MacKay 1991, 1992b). The results in this chapter clarify the meaning of this procedure — by allowing σ_u to be set by the data, we let the data determine the scale above which the function takes on a Brownian character (see equation (2.11)). The results concerning fractional Brownian priors suggest that it might be useful to make η a hyperparameter as well, to allow the fractional Brownian character of the function to be determined by the data. Similarly, when using a t-distribution as a prior for weights, it might be useful to make the shape parameter, α, be a hyperparameter, and thereby allow the index of the stable distribution to which the prior converges to vary.

Consideration of the results in the chapter also reveals a potential problem when these hierarchical models are used with networks having large numbers of hidden units. The extent of the central region over which the characteristics of functions drawn from the prior are approximately uniform is determined by the ratio σ_a/σ_u. When these quantities are hyperparameters, the size of this region can vary independently of the smoothness characteristics of the function, which depend only on σ_u. Typically, the size of this region will not be fixed by the data — if the data indicate that the properties of the actual function are uniform over the region for which training data is available, then *any* values of the hyperparameters that lead to a central region at least this large will be compatible with the data. If the number of hidden units is small, the central region will presumably be forced to have approximately the same extent as the training data, in order that all the hidden units can be exploited. When there are many hidden units, however, the pressure for them to all be used to explain the training data will be much less, and the size of the central region will be only loosely constrained.

This phenomenon will not necessarily lead to bad predictive performance — indeed, if extrapolation outside the region of the training data is to be done, it is desirable for the central region to extend beyond the training data, to include the region where predictions are to be made. If we are interested only in the training region, however, using a model whose central region is much larger then the training region may lead to substantial wasted computation, as many hidden units in the network will have no influence on the area of interest. Some reformulation of the model that allowed the user to exercise greater control over the central region would be of interest.

Chapter 3
Monte Carlo Implementation

This chapter presents a Markov chain Monte Carlo implementation of Bayesian learning for neural networks in which network parameters are updated using the hybrid Monte Carlo algorithm, a form of the Metropolis algorithm in which candidate states are found by means of dynamical simulation. Hyperparameters are updated separately using Gibbs sampling, allowing their values to be used in chosing good stepsizes for the discretized dynamics. I show that hybrid Monte Carlo performs better than simple Metropolis, due to its avoidance of random walk behaviour. I also discuss variants of hybrid Monte Carlo in which dynamical computations are done using "partial gradients", in which acceptance is based on a "window" of states, and in which momentum updates incorporate "persistence".

The implementation of Bayesian learning for multilayer perceptron networks due to MacKay (1991, 1992b) uses a Gaussian approximation for the posterior distribution of the network parameters (weights and biases), and single-valued estimates for the hyperparameters (prior variances for the parameters, and the noise variance). Such approximate Bayesian methods have proven useful in some practical applications (MacKay 1994a, Thodberg 1996). However, as discussed in Chapter 1, there are reasons to believe that these methods will not always produce good approximations to the true result implied by the model, especially if complex models are used in order to take full advantage of the available data.

There is thus a need for an implementation of Bayesian learning that does not rely on any assumptions concerning the form of the posterior distribution. At a minimum, such an implementation would be useful in assessing the accuracy of methods based on Gaussian approximations. If Gaussian approximation methods are often inadequate, as I expect, an implementation that avoids such assumptions will be necessary in order to assess the true merits of Bayesian neural network models, and to apply them with confidence in practical situations.

Bayesian learning for neural networks is a difficult problem, due to the typically complex nature of the posterior distribution. At present, it appears that only Markov chain Monte Carlo methods (reviewed in Section 1.3.1) offer any hope of producing in a feasible amount of time results whose accuracy is reasonably assured, without the need for any questionable assumptions. As will be seen, however, the Markov chain Monte Carlo methods commonly used for statistical applications are either not applicable to this problem or are very slow. Better results can be obtained using the hybrid Monte Carlo algorithm, due to its avoidance of random walk behaviour. Hybrid Monte Carlo was originally developed for use in quantum chromodynamics, and is not widely known outside the lattice field theory community. I believe this algorithm is of general interest, however, and will prove useful in many statistical applications.

I begin this chapter by reviewing the hybrid Monte Carlo algorithm, after which I describe an implementation of Bayesian learning for multilayer perceptron networks based on it. The range of network models handled by this implementation and the details of the computational methods used are described in Appendix A. I demonstrate the use of this implementation on the "robot arm" problem of MacKay (1991, 1992b). I then compare the performance of hybrid Monte Carlo with other methods, such as simple forms of the Metropolis algorithm. I conclude by examining several variants of the basic hybrid Monte Carlo method, which can sometime improve performance.

Note that throughout this chapter the objective is to develop a computationally feasible procedure for producing the Bayesian predictions that are mathematically implied by the model being employed. Whether such predictions are good, in the sense of being close to the true values, is another matter, consideration of which is for the most part deferred to Chapter 4. The use of this implementation in Chapter 4 will also further test its computational performance, for a variety of networks architectures, data models, and priors.

3.1 The hybrid Monte Carlo algorithm

The *hybrid Monte Carlo* algorithm of Duane, Kennedy, Pendleton, and Roweth (1987) merges the Metropolis algorithm with sampling techniques based on dynamical simulation. The output of the algorithm is a sample of points drawn from some specified distribution, which can then be used to form Monte Carlo estimates for the expectations of various functions with respect to this distribution (see equation (1.14)). For Bayesian learning, we wish to sample from the posterior distribution given the training data, and are interested in estimating the expectations needed to make predictions for test cases, such as in equation (1.12).

One way of viewing the hybrid Monte Carlo algorithm is as a combination of Gibbs sampling and a particularly elaborate version of the Metropolis algorithm. I assume here that the reader is familiar with these two methods, which were reviewed in Section 1.3.1. The hybrid Monte Carlo algorithm itself, and methods related to it, have been reviewed by Toussaint (1989), Kennedy (1990), and myself (Neal 1993b).

3.1.1 *Formulating the problem in terms of energy*

The hybrid Monte Carlo algorithm is expressed in terms of sampling from the *canonical* (or *Boltzmann*) distribution for the state of a physical system, which is defined in terms of an energy function. However, the algorithm can be used to sample from any distribution for a set of real-valued variables for which the derivatives of the probability density can be computed. It is convenient to retain the physical terminology even in non-physical contexts, by formulating the problem in terms of an energy function for a fictitious physical system.

Accordingly, suppose we wish to sample from some distribution for a "position" variable, q, which has n real-valued components, q_i. In a real physical system, q would consist of the coordinates of all the particles; in our application, q will be the set of network parameters. The probability density for this variable under the canonical distribution is defined by

$$P(q) \;\propto\; \exp(-E(q)) \qquad (3.1)$$

where $E(q)$ is the "potential energy" function. (The "temperature" parameter of the canonical distribution is here set to one, as it plays no role in the present application.) Any probability density that is nowhere zero can be put in this form, by simply defining $E(q) = -\log P(q) - \log Z$, for any convenient Z.

To allow the use of dynamical methods, we introduce a "momentum" variable, p, which has n real-valued components, p_i, in one-to-one correspondence with the components of q. The canonical distribution over the "phase space" of q and p together is defined to be

$$P(q, p) \;\propto\; \exp(-H(q, p)) \qquad (3.2)$$

where $H(q,p) = E(q) + K(p)$ is the "Hamiltonian" function, which gives the total energy. $K(p)$ is the "kinetic energy" due to the momentum, for which the usual choice is

$$K(p) = \sum_{i=1}^{n} \frac{p_i^2}{2m_i} \qquad (3.3)$$

The m_i are the "masses" associated with each component. Adjustment of these mass values can improve efficiency, but for the moment they may be taken to all be one.

In the distribution of equation (3.2), q and p are independent, and the marginal distribution of q is the same as that of equation (3.1), from which we wish to sample. We can therefore proceed by defining a Markov chain that converges to the canonical distribution for q and p, and then simply ignore the p values when estimating expectations of functions of q. This manoeuver may appear pointless at present, but will eventually be shown to yield substantial benefits through its suppression of random walk behaviour.

3.1.2 The stochastic dynamics method

Hybrid Monte Carlo can be viewed as an elaboration of the *stochastic dynamics* method (Andersen 1980), in which the task of sampling from the canonical distribution for q and p given by equation (3.2) is split into two sub-tasks — sampling uniformly from values of q and p with a fixed total energy, $H(q,p)$, and sampling states with different values of H.

Sampling at a fixed total energy is done by simulating the *Hamiltonian dynamics* of the system, in which the state evolves in fictitious time, τ, according to the following equations:

$$\frac{dq_i}{d\tau} = +\frac{\partial H}{\partial p_i} = \frac{p_i}{m_i} \qquad (3.4)$$

$$\frac{dp_i}{d\tau} = -\frac{\partial H}{\partial q_i} = -\frac{\partial E}{\partial q_i} \qquad (3.5)$$

To do this, we must be able to compute the partial derivatives of E with respect to the q_i.

Three properties of Hamiltonian dynamics are crucial to its use in sampling. First, H stays constant as q and p vary according to this dynamics, as can be seen as follows:

$$\frac{dH}{d\tau} = \sum_i \left[\frac{\partial H}{\partial q_i} \frac{dq_i}{d\tau} + \frac{\partial H}{\partial p_i} \frac{dp_i}{d\tau} \right] \qquad (3.6)$$

$$= \sum_i \left[\frac{\partial H}{\partial q_i} \frac{\partial H}{\partial p_i} - \frac{\partial H}{\partial p_i} \frac{\partial H}{\partial q_i} \right] = 0 \qquad (3.7)$$

Second, Hamiltonian dynamics preserves the volumes of regions of phase space — i.e. if we follow how the points in some region of volume V move according to the dynamical equations, we find that the region where these points end up after some given period of time also has volume V. We can see this by looking at the divergence of the motion in phase space:

$$\sum_i \left[\frac{\partial}{\partial q_i}\left(\frac{dq_i}{d\tau}\right) + \frac{\partial}{\partial p_i}\left(\frac{dp_i}{d\tau}\right) \right] = \sum_i \left[\frac{\partial H}{\partial q_i \partial p_i} - \frac{\partial H}{\partial p_i \partial q_i} \right] = 0 \quad (3.8)$$

Finally, the dynamics is reversible. After following the dynamics forward in time for some period, we can recover the original state by following the dynamics backward in time for an equal period. We cann also return to the initial state by negating the momentum variables, following the dynamics for the same period, and then negating the momentum variables again.

Together, these properties imply that the canonical distribution for q and p is invariant with respect to transitions that consist of following a trajectory for some pre-specified period of time using Hamiltonian dynamics. The probability that we will end in some small region after the transition will be the same as the probability that we started in the corresponding region (of equal volume) found by reversing the dynamics. If this probability is given by the canonical distribution, the probability of being in the final region will also be in accord with the canonical distribution, since the probabilities under the canonical distribution depend only on H, which is the same at the start and end of the trajectory.

In many cases, transitions based on Hamiltonian dynamics will eventually explore the whole region of phase space with a given value of H. Such transitions are clearly not sufficient to produce an ergodic Markov chain, however, since regions with different values of H are never visited.

In the stochastic dynamics method, an ergodic Markov chain is obtained by alternately performing deterministic dynamical transitions and stochastic Gibbs sampling ("heatbath") updates of the momentum. Since q and p are independent, p may be updated without reference to q by drawing a new value with probability density proportional to $\exp(-K(p))$. For the kinetic energy function of equation (3.3), this is easily done, since the p_i have independent Gaussian distributions. These updates of p can change H, allowing the entire phase space to be explored.

The length in fictitious time of the trajectories is an adjustable parameter of the stochastic dynamics method. It is best to use trajectories that result in large changes to q. This avoids the random walk effects that would result from randomizing the momentum after every short trajectory. (This point is discussed further below, in connection with hybrid Monte Carlo.)

In practice, Hamiltonian dynamics cannot be simulated exactly, but can only be approximated by some discretization using finite time steps. In the *leapfrog* discretization, a single step finds approximations to the position and momentum, $\widehat{q}$ and $\widehat{p}$, at time $\tau + \epsilon$ from $\widehat{q}$ and $\widehat{p}$ at time τ as follows:

$$\widehat{p}_i(\tau + \tfrac{\epsilon}{2}) = \widehat{p}_i(\tau) - \frac{\epsilon}{2}\frac{\partial E}{\partial q_i}(\widehat{q}(\tau)) \tag{3.9}$$

$$\widehat{q}_i(\tau + \epsilon) = \widehat{q}_i(\tau) + \epsilon \frac{\widehat{p}_i(\tau + \tfrac{\epsilon}{2})}{m_i} \tag{3.10}$$

$$\widehat{p}_i(\tau + \epsilon) = \widehat{p}_i(\tau + \tfrac{\epsilon}{2}) - \frac{\epsilon}{2}\frac{\partial E}{\partial q_i}(\widehat{q}(\tau + \epsilon)) \tag{3.11}$$

Such a leapfrog step consists of a half-step for the p_i, a full step for the q_i, and another half-step for the p_i. (One can instead do a half-step for the q_i, a full step for the p_i, and another half-step for the q_i, but this is usually slightly less convenient.) To follow the dynamics for some period of time, $\Delta\tau$, a value is chosen for the stepsize, ϵ, that is thought to be small enough to give acceptable error, and equations (3.9)–(3.11) are applied for $L = \Delta\tau/\epsilon$ steps in order to reach the target time. When this is done, the last half-step for p_i in one leapfrog step will be immediately followed by the first half-step for p_i in the next leapfrog step. All but the very first and very last such half-steps can therefore be merged into full steps starting at times $\tau + k\epsilon + \epsilon/2$, which "leapfrog" over the steps for the q_i that start at times $\tau + k\epsilon$.

In the leapfrog discretization, phase space volume is still preserved (a consequence of the fact that each of the changes to a component of q or p in a leapfrog step depends only on the current values of the *other* components). The dynamics can also still be reversed (by simply applying the same number of leapfrog steps with ϵ negated). However, the value of H no longer stays exactly constant. Because of this, Monte Carlo estimates found using the stochastic dynamics method will suffer from some systematic error, which will go to zero only as the stepsize, ϵ, is reduced to zero (with the number of steps needed to compute each trajectory then going to infinity).

3.1.3 Hybrid Monte Carlo

In the hybrid Monte Carlo algorithm of Duane, *et al* (1987), the systematic error of the stochastic dynamics method is eliminated by merging it with the Metropolis algorithm.

Like the uncorrected stochastic dynamics method, the hybrid Monte Carlo algorithm samples points in phase space by means of a Markov chain in which stochastic and dynamical transitions alternate. In the stochastic transitions, the momentum is replaced using Gibbs sampling, as described in the previous section. The dynamical transitions in the hybrid Monte Carlo method are also similar to those in the stochastic dynamics method,

3.1 The hybrid Monte Carlo algorithm

but with two changes — first, the momentum is negated after the trajectory is computed; second, the point reached by following the dynamics is only a candidate for the new state, to be accepted or rejected based on the change in total energy, as in the Metropolis algorithm. If the dynamics were simulated exactly, the change in H would always be zero, and the new point would always be accepted. When the dynamics is simulated using some approximate discretization, H may change, and moves will occasionally be rejected. These rejections exactly eliminate the bias introduced by the inexact simulation.

In detail, given values for the leapfrog stepsize, ϵ, and the number of leapfrog steps, L, a dynamical transition is performed as follows:

1) Starting from the current state, $(q,p) = (\widehat{q}(0), \widehat{p}(0))$, perform L leapfrog steps with a stepsize of ϵ, resulting in the state $(\widehat{q}(\epsilon L), \widehat{p}(\epsilon L))$.

2) Negate the momentum variables, thereby producing the state $(q^*, p^*) = (\widehat{q}(\epsilon L), -\widehat{p}(\epsilon L))$.

3) Regard (q^*, p^*) as a candidate for the next state, as in the Metropolis algorithm, accepting it with probability
$$\min\left(1,\, \exp\left(-(H(q^*, p^*) - H(q, p))\right)\right),$$
and otherwise letting the new state be the same as the old.

The negation of the momentum in step (2), together with the reversibility of the leapfrog dynamics, ensures that if we were to perform a dynamical transition starting with the candidate state above, (q^*, p^*), the state proposed would be the initial state above, (q, p). Furthermore, since the leapfrog steps preserve phase space volume, points in phase space are not squeezed together or spread apart by the mapping from current to proposed state. The proposal of a candidate state above therefore has the symmetry required for a Metropolis update to leave the desired distribution invariant (see Section 1.3.3). The negation of the momentum variables in step (2) is in fact unnecessary if the momentum will be replaced in a Gibbs sampling step before the next dynamical transition anyway, but it is necessary if the dynamical transitions are employed in some other context, as in Section 3.5.3.

The values for ϵ and for L used above may be chosen at random from some fixed distribution. This may be useful when the best values are not known, or vary from place to place. Some random variation may also be needed to avoid periodicities that could interfere with ergodicity (Mackenzie 1989), though this is not expected to be a problem for an irregular distribution such as a neural network posterior.

The name *Langevin Monte Carlo* is given to hybrid Monte Carlo with $L = 1$, that is, in which candidate states are generated using only a single leapfrog step. The "smart Monte Carlo" method of Rossky, Doll, and Friedman (1978) is equivalent to this.

Only when L is reasonably large, however, does one obtain the principal benefit of hybrid Monte Carlo — the avoidance of random walks. One might think that a large error in H would develop over a long trajectory, leading to a very low acceptance rate. For sufficiently small stepsizes, this usually does not occur. Instead, the value of H oscillates along the trajectory, and the acceptance rate is almost independent of trajectory length. For stepsizes above a certain limit, however, the leapfrog discretization becomes unstable, and the acceptance rate is very low. The optimal strategy is usually to select a stepsize just a bit below this point of instability. Trajectories should be made long enough that they typically lead to states distant from their starting point, but no longer. Shorter trajectories would result in the distribution being explored via a random walk; longer trajectories would wastefully traverse the whole distribution several times, ending finally at a point similar to one that might have been reached by a shorter trajectory.

Figure 3.1 illustrates the advantage of using long trajectories in hybrid Monte Carlo. Here, the distribution for $q = (q_1, q_2)$ that we wish to sample from is a bivariate Gaussian with high correlation, defined by the potential energy function

$$E(q) = \left(q_1^2/\sigma_1^2 + q_2^2/\sigma_2^2 - 2\rho q_1 q_2/\sigma_1\sigma_2 \right) \Big/ 2(1-\rho^2) \quad (3.12)$$

We could of course transform to a different coordinate system in which the two components are independent, at which point sampling would become easy. In more complex problems this will be difficult, however, so we assume that we cannot do this. If the masses, m_1 and m_2, associated with the two components are set to one, the leapfrog method is stable for this problem as long as the stepsize used is less than twice the standard deviation in the most confined direction; to keep the rejection rate low, we will have to limit ourselves to a stepsize a bit less than this. Many leapfrog steps will therefore be needed to explore in the less confined direction.

The left of Figure 3.1 shows the progress of twenty Langevin Monte Carlo iterations. In each iteration, the momentum is replaced from its canonical distribution, and a single leapfrog step is then performed (with the result sometimes being rejected). Due to the randomization of the direction each iteration, the progress takes the form of a random walk. If each iteration moves a distance of about ℓ, then k iterations will typically move a distance of only about $\ell\sqrt{k}$.

The right the Figure 3.1 shows a single hybrid Monte Carlo trajectory consisting of twenty leapfrog steps, with the momentum being randomized only at the start. Such trajectories move consistently in one direction, until they are "reflected" upon reaching a region of low probability. Accordingly, in k steps that each move a distance of about ℓ, the hybrid Monte Carlo

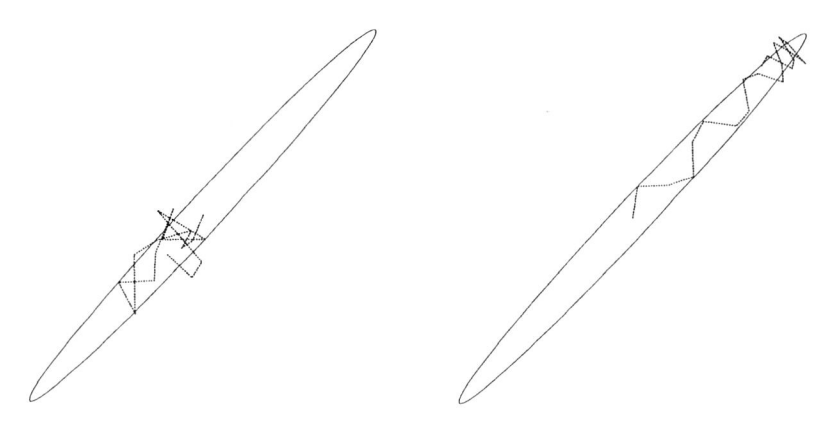

FIGURE 3.1. Sampling using the Langevin and hybrid Monte Carlo methods. The distribution sampled from is a bivariate Gaussian with $\sigma_1 = \sigma_2 = 1$, and $\rho = 0.99$, represented above by its one standard deviation contour. Sampling by Langevin Monte Carlo is illustrated on the left, which shows twenty single-step trajectories (except some rejected trajectories are not shown). Sampling by hybrid Monte Carlo is illustrated on the right, which shows a single twenty-step trajectory. In both cases, the leapfrog method was used with a stepsize of 0.15. Only the course of the position variables is depicted; the momentum variables are not shown.

can move a distance of up to ℓk, permitting much more efficient exploration than is obtained with a random walk.

The Langevin Monte Carlo method does permit use of a somewhat larger leapfrog stepsize while maintaining a good acceptance rate, but for distributions with high correlations this advantage is more than offset by the penalty from performing a random walk. Gibbs sampling for such distributions also produces a random walk, with similar size changes. In a simple version of the Metropolis algorithm, in which candidate states are drawn from a symmetric Gaussian distribution centred at the current point, maintaining a high acceptance rate requires limiting the size of the changes to about the same amount as are produced with Langevin Monte Carlo or Gibbs sampling, again resulting in a random walk. (For this two-dimensional problem, simple Metropolis in fact performs best when quite large changes are proposed, even though the acceptance rate is then very low, but this strategy ceases to work in higher-dimensional problems.)

3.2 An implementation of Bayesian neural network learning

Bayesian learning and its application to multilayer perceptron networks were discussed in Chapter 1. I will recap the notation here. The network is parameterized by weights and biases, collectively denoted by θ, that define what function from inputs to outputs is computed by the network. This function is written as $f(x, \theta)$. A prior for the network parameters is defined, which may depend on the values of some hyperparameters, γ. The prior density for the parameters is written as $P(\theta \mid \gamma)$, the prior density for the hyperparameters themselves as $P(\gamma)$. We have a set of training cases, $(x^{(1)}, y^{(1)}), \ldots, (x^{(n)}, y^{(n)})$, consisting of independent pairs of input values, $x^{(i)}$, and target values, $y^{(i)}$. We aim to model the conditional distribution for the target values given the input values, which we specify in terms of $f(x, \theta)$, perhaps also using the hyperparameters, γ. These conditional probabilities or probability densities for the target are written as $P(y \mid x, \theta, \gamma)$.

Our ultimate objective is to predict the target value for a new test case, $y^{(n+1)}$, given the corresponding inputs, $x^{(n+1)}$, using the information in the training set. This prediction is based on the posterior distribution for θ and γ, which is proportional to the product of the prior and the likelihood due to the training cases:

$$P(\theta, \gamma \mid (x^{(1)}, y^{(1)}), \ldots, (x^{(n)}, y^{(n)}))$$

$$\propto\; P(\gamma)\, P(\theta \mid \gamma) \prod_{c=1}^{n} P(y^{(c)} \mid x^{(c)}, \theta, \gamma) \qquad (3.13)$$

Predictions are made by integration with respect to this posterior distribution. The full predictive distribution is

$$P(y^{(n+1)} \mid x^{(n+1)}, (x^{(1)}, y^{(1)}), \ldots, (x^{(n)}, y^{(n)})) \qquad (3.14)$$

$$= \int P(y^{(n+1)} \mid x^{(n+1)}, \theta, \gamma)\, P(\theta, \gamma \mid (x^{(1)}, y^{(1)}), \ldots, (x^{(n)}, y^{(n)}))\, d\theta\, d\gamma$$

For a regression model, the single-valued prediction that minimizes expected squared-error loss is the mean of the predictive distribution. If the conditional distribution for the targets is defined to have a mean given by the corresponding network outputs, this optimal prediction is

$$\widehat{y}^{(n+1)} \;=\; \int f(x^{(n+1)}, \theta)\, P(\theta, \gamma \mid (x^{(1)}, y^{(1)}), \ldots, (x^{(n)}, y^{(n)}))\, d\theta\, d\gamma \quad (3.15)$$

In the Markov chain Monte Carlo approach, these integrals, which take the form of expectations of functions with respect to the posterior distribution, are approximated by the average value of the function over a sample of values from the posterior.

I believe that hybrid Monte Carlo is the most promising Markov chain method for sampling from the posterior distribution of a neural network model. One cannot even attempt to use ordinary Gibbs sampling for this problem, since sampling from the conditional distributions is infeasible.

3.2 An implementation of Bayesian neural network learning 65

Simple forms of the Metropolis algorithm are possible, but will suffer from random walks. Uncorrected stochastic dynamics (see Section 3.1.2) can also be applied to this problem (Neal 1993a), but as this raises the possibility of unrecognized systematic error, the hybrid Monte Carlo method appears to be the safer choice. These other methods will be compared to hybrid Monte Carlo in Section 3.4.

There are many possible ways of using hybrid Monte Carlo to sample from the posterior distribution for a neural network model. In my earliest work on this problem (Neal 1992b), I felt that use of "simulated annealing" (Kirkpatrick, Gelatt, and Vecchi 1983) was desirable, in order to overcome the potential problem that the simulation could be trapped for a long time in a local minimum of the energy. I therefore chose a parameterization of the model in which the prior was uniform, since this allows annealing to be done without affecting the prior. In the simulation results I reported, annealing was indeed found to be beneficial. However, later work revealed that the primary benefit of annealing was in overcoming the bad effects of the parameterization used — which had been chosen only because it made annealing more convenient!

In later work, I therefore abandoned use of annealing (though it remains possible that it might be beneficial in some form). Many other implementation decisions remain, however.

Hyperparameters can be handled in several ways. In previous implementations (Neal 1992a, 1993a), I replaced them with equivalent scale factors. Rather than letting the standard deviation of a group of weights, w_i, be controlled by a hyperparameter, σ, I instead expressed these weights in terms of a scale factor, s, and a set of unscaled weights, u_i, with $w_i = su_i$. The prior distribution for the u_i was fixed, with a standard deviation of one, while s was given its own prior. Hybrid Monte Carlo was then applied to update both s and the u_i. While this method worked reasonably well, it had the undesirable effect that the optimal stepsize for use with the u_i would vary with the current value of s.

The choices made in the implementation described in this chapter are based in part on this previous experience. I cannot claim that my latest choices are optimal, however. Many possibilities remain to be evaluated, and I expect that the performance reported here may ultimately be improved upon significantly.

I aim in this implementation to handle a wide range of network architectures and associated data models. Both regression and classification models are implemented, networks with any number of hidden layers are allowed, and prior distributions that include all those discussed in Chapter 2 are supported (except for those based on step-function hidden units, which are not suitable for implementation using backpropagation). Not all aspects of these models are discussed in detail here, but they are described in Appendix A. Many useful extensions have not yet been implemented, but could be within the general framework of this implementation. Such possible extensions include those mentioned in Section 2.4 in which η (controlling the fractional Brownian character of the function) and α (controlling the index of the stable distribution) are treated as hyperparameters, and regression models in which the noise level varies depending on the inputs ("heteroscedasticity", in statistical parlance).

Another objective of this implementation is to minimize the amount of "tuning" that is needed to obtain good performance. Gibbs sampling is very nice in this respect, as it has no tunable parameters. In simple forms of the Metropolis algorithm, one must decide on the magnitude of the changes proposed, and in hybrid Monte Carlo one must select both the stepsize, ϵ, and the number of leapfrog steps, L. I attempt in this implementation to derive the stepsizes automatically, though the user must still adjust these stepsizes by a small amount to get good performance. Specifying the number of leapfrog steps in a trajectory is still left to the user.

The scheme used for setting stepsizes relies on a separation of the updates for the hyperparameters from the updates for the network parameters (weights and biases). The hyperparameters are updated by Gibbs sampling. The network parameters are updated by hybrid Monte Carlo, using stepsizes that depend on the current values of the hyperparameters. These two aspects of the implementation will now be described in turn.

3.2.1 Gibbs sampling for hyperparameters

Two types of hyperparameters are present in neural network models — those in terms of which the prior distribution of the parameters is expressed, and those that specify the noise levels in regression models. One might not regard quantities of the latter type as "hyperparameters", since they do not control the distribution of lower-level "parameters", but I use the same term here because in this implementation quantities of both types are handled similarly, via Gibbs sampling. These quantities also handled similarly in the implementation of MacKay (1991, 1992b) and in the work of Buntine and Weigend (1991).

In the simplest cases, a hyperparameter of the first type controls the standard deviation for all parameters in a certain group. Such a group might consist of the biases for all units of one type, or the weights on all connections from units of one type to those of another type, or the weights on all connections from a particular unit to units of some type. The manner in which parameters are grouped is a modeling choice that is made on the basis of prior knowledge.

3.2 An implementation of Bayesian neural network learning

In detail, let the parameters in a particular group be $u_1, \ldots, u_k$ (in the notation given previously, these are components of θ). Conditional on the value of the controlling hyperparameter, let the parameters in this group be independent, and have Gaussian distributions with mean zero and standard deviation σ_u. It is convenient to represent this standard deviation in terms of the corresponding "precision", defined to be $\tau_u = \sigma_u^{-2}$. The distribution for the parameters in the group is then given by

$$P(u_1, \ldots, u_k \mid \tau_u) \;=\; (2\pi)^{-k/2} \tau_u^{k/2} \exp\left(-\tau_u \sum_i u_i^2 / 2\right) \quad (3.16)$$

The precision is given a Gamma distribution with some mean, ω_u, and shape parameter specified by α_u, with density

$$P(\tau_u) \;=\; \frac{(\alpha_u/2\omega_u)^{\alpha_u/2}}{\Gamma(\alpha_u/2)} \tau_u^{\alpha_u/2 - 1} \exp\left(-\tau_u \alpha_u / 2\omega_u\right) \quad (3.17)$$

In the previous notation, τ_u is a component of γ. The values of ω_u and α_u may for the moment be considered fixed.

The prior for τ_u is "conjugate" to its use in defining the distribution for the u_i. The conditional distribution for τ_u given values for the u_i is therefore also of the Gamma form:

$$P(\tau_u \mid u_1, \ldots, u_k)$$
$$\propto \; \tau_u^{\alpha_u/2 - 1} \exp(-\tau_u \alpha_u / 2\omega_u) \cdot \tau_u^{k/2} \exp\left(-\tau_u \sum_i u_i^2 / 2\right) \quad (3.18)$$
$$\propto \; \tau_u^{(\alpha_u + k)/2 - 1} \exp\left(-\tau_u (\alpha_u/\omega_u + \sum_i u_i^2) / 2\right) \quad (3.19)$$

From the above expression, one can see that the prior for τ_u can be interpreted as specifying α_u imaginary parameter values, whose average squared magnitude is $1/\omega_u$. Small values of α_u produce vague priors for τ_u.

The conditional distribution of equation (3.19) is what is needed for a Gibbs sampling update, since given $u_1, \ldots, u_k$, the value of τ_u is independent of the other parameters, hyperparameters, and target values. Efficient methods of generating Gamma-distributed random variates are known (Devroye 1986).

The implementation described in Appendix A allows for more complex situations, in which the priors for the precisions may be specified using higher-level hyperparameters. For example, each hidden unit might have an associated hyperparameter giving the precision for weights out of that unit, with the mean for these precisions (ω in equation (3.17)) being a common higher-level hyperparameter, shared by all units of one type. Gibbs sampling for the lower-level hyperparameters remains as above, but more complex methods are needed to implement Gibbs sampling for the higher-level hyperparameter. The distribution given to a single parameter may also be a t-distribution, rather than a Gaussian. Since t-distributions can be represented as mixtures of Gaussian distributions with precisions given by Gamma distributions, this can be implemented by extending the hierarchy downward, to include implicit precision variables associated with individual parameters.

The treatment of hyperparameters specifying the amount of noise in regression models is similar. Again, it is convenient for the hyperparameters to be precision values, $\tau_k = \sigma_k^{-2}$, where σ_k is here the standard deviation of the noise associated with the kth target value. Given the inputs, network parameters, and the noise standard deviations, the various target values in the training set are independent, giving:

$$P(y_k^{(1)}, \ldots, y_k^{(n)} \mid x^{(1)}, \ldots, x^{(n)}, \theta, \tau_k)$$
$$= (2\pi)^{-n/2} \tau_k^{n/2} \exp\left(-\tau_k \sum_c \left(y_k^{(c)} - f_k(x^{(c)}, \theta)\right)^2 / 2\right) \quad (3.20)$$

As before, we give τ_k a Gamma prior:

$$P(\tau_k) = \frac{(\alpha/2\omega)^{\alpha/2}}{\Gamma(\alpha/2)} \tau_k^{\alpha/2-1} \exp\left(-\tau_k \alpha / 2\omega\right) \quad (3.21)$$

and obtain a Gamma distribution for τ_k given everything else:

$$P(\tau_k \mid (x^{(1)}, y^{(1)}), \ldots, (x^{(n)}, y^{(n)}), \theta)$$
$$\propto \tau_k^{(\alpha+n)/2-1} \exp\left(-\tau_k(\alpha/\omega + \sum_c \left(y_k^{(c)} - f_k(x^{(c)}, \theta)\right)^2 / 2\right) \quad (3.22)$$

Variations on this scheme described in Appendix A include models with higher-level hyperparameters linking the τ_k, or alternatively that use a single τ for all targets, and models in which the noise follows a t-distribution rather than a Gaussian.

3.2.2 Hybrid Monte Carlo for network parameters

A Markov chain that explores the entire posterior distribution can be obtained by alternating Gibbs sampling updates for the hyperparameters, as described in the previous section, with hybrid Monte Carlo updates for the network parameters.

To apply the hybrid Monte Carlo method, we must formulate the desired distribution in terms of a potential energy function. Since we wish to sample from the posterior distribution for network parameters (the weights and biases), the energy will be a function of these parameters, previously called θ, which now play the role of the "position" variables, q, of an imaginary physical system. (From here on, θ and q will be used interchangeably, as appropriate in context). The hyperparameters will remain fixed throughout one hybrid Monte Carlo update, so we can omit from the energy any terms that depend only on the hyperparameters. For the generic case described by equation (3.13), the potential energy is derived from the log of the prior and the log of the likelihood due to the training cases, as follows:

$$E(\theta) = F(\gamma) - \log P(\theta \mid \gamma) - \sum_{c=1}^{n} \log P(y^{(c)} \mid x^{(c)}, \theta, \gamma) \quad (3.23)$$

where $F(\gamma)$ is any function of of the hyperparameters that we find convenient. The canonical distribution for this energy function, which is proportional to $\exp(-E(\theta))$, will then produce the posterior probability density for θ given γ. Note that the energy function will change whenever the hyperparameters change, which will normally be between successive hybrid Monte Carlo updates, when new values for the hyperparameters are chosen using Gibbs sampling.

The detailed form of the energy function will vary with the network architecture, the prior, and the data model used. As a specific example, suppose that the network parameters form two groups, u and v, so that $\theta = \{u_1, \ldots, u_k, v_1, \ldots, v_h\}$; let the prior standard deviations for these two groups be σ_u and σ_v. Suppose also that the target is a single real value, modeled with a Gaussian noise distribution of standard deviation σ. The hyperparameters are then $\gamma = \{\tau_u, \tau_v, \tau\}$, where $\tau_u = \sigma_u^{-2}$, $\tau_v = \sigma_v^{-2}$, and $\tau = \sigma^{-2}$. The priors for the two groups of weights conditional on the hyperparameters are of the form given by equation (3.16), and the likelihood due to the training cases is given by equation (3.20). The resulting potential energy function is

$$E(\theta) = \tau_u \sum_{i=1}^{k} u_i^2/2 + \tau_v \sum_{j=1}^{h} v_j^2/2 + \tau \sum_{c=1}^{n} (y^{(c)} - f(x^{(c)}, \theta))^2/2 \quad (3.24)$$

It is helpful to impose a very large upper limit (e.g. 10^{30}) on the value of E above. This avoids problems with floating-point overflow during computation of trajectories that turn out to be unstable, since the derivatives of E at points where the limit is exceeded are zero, preventing the instability from going further.

This energy function is similar to the error function (with weight decay penalty) that is minimized in conventional neural network training. Recall, however, that the objective in a Monte Carlo implementation of Bayesian learning is not to find the minimum of the energy, but rather to sample from the corresponding canonical distribution.

To sample from this canonical distribution using the hybrid Monte Carlo method, we introduce momentum variables, p_i, in one-to-one correspondence with the position variables, q_i, which are here identified with the parameters, θ. With each momentum variable, we also associate a positive "mass", m_i. These masses are used in defining the kinetic energy, $K(p)$, associated with the momentum (equation (3.3)), with the result that the canonical distributions for the p_i are Gaussian with means of zero and variances m_i (independently of each other and of the position). As described in Section 3.1.3, a single hybrid Monte Carlo update starts by generating new values for all the momentum variables from their canonical distribution. A candidate state is then found by following a trajectory computed using the leapfrog discretization of Hamiltonian dynamics (equations (3.9)–(3.11)), applied for some number of steps, L, using some stepsize, ϵ. Finally this candidate is accepted or rejected based on the change in total energy, $H(q,p) = E(q) + K(p)$. Calculation of the derivatives of E with respect to the q_i is required in order to perform the leapfrog steps; these derivatives can be found by the usual "backpropagation" method (Rumelhart, Hinton, and Williams 1986a, 1986b).

We would like to set the masses, m_i, the stepsize, ϵ, and the number of leapfrog steps in a trajectory, L, to values that will produce a Markov chain that converges rapidly to the posterior distribution, and then rapidly moves about the posterior. Rapid movement will keep the dependence between states in the Markov chain low, which typically increases the accuracy of Monte Carlo estimates based on a given number of such states (see Section 1.3.1). In this implementation, the number of leapfrog steps must be set by the user. (Ways of making this choice are discussed in connection with the demonstration of Section 3.3.) I attempt to set the masses and the stepsize automatically, but the user may still need to adjust these quantities based on the observed rejection rate.

It is convenient to recast the choice of masses, m_i, and stepsize, ϵ, as a choice of individual stepsizes, ϵ_i, that are applied when updating each component of the position and momentum. The leapfrog method of equations (3.9)–(3.11) can be rewritten as follows:

$$\frac{\widehat{p}_i(\tau + \frac{\epsilon}{2})}{\sqrt{m_i}} = \frac{\widehat{p}_i(\tau)}{\sqrt{m_i}} - \frac{\epsilon/\sqrt{m_i}}{2} \frac{\partial E}{\partial q_i}(\widehat{q}(\tau)) \quad (3.25)$$

$$\widehat{q}_i(\tau + \epsilon) = \widehat{q}_i(\tau) + (\epsilon/\sqrt{m_i}) \frac{\widehat{p}_i(\tau + \frac{\epsilon}{2})}{\sqrt{m_i}} \quad (3.26)$$

$$\frac{\widehat{p}_i(\tau + \epsilon)}{\sqrt{m_i}} = \frac{\widehat{p}_i(\tau + \frac{\epsilon}{2})}{\sqrt{m_i}} - \frac{\epsilon/\sqrt{m_i}}{2} \frac{\partial E}{\partial q_i}(\widehat{q}(\tau + \epsilon)) \quad (3.27)$$

Rather than applying the leapfrog equations to update p_i and q_i, we can therefore store the values $p_i/\sqrt{m_i}$ instead of the p_i, and update these values (along with the q_i) using leapfrog steps in which different components have different stepsizes, given by $\epsilon_i = \epsilon/\sqrt{m_i}$.

3.2 An implementation of Bayesian neural network learning

This re-expression of the leapfrog method reduces slightly the amount of computation required, and has the additional advantage that the canonical distribution of $p_i/\sqrt{m_i}$ is independent of m_i. Accordingly, after a change in the m_i, the $p_i/\sqrt{m_i}$ values will be distributed according to the new canonical distribution as long as they were previously distributed according to the old canonical distribution. In this implementation, the m_i (equivalently, the ϵ_i) are set based on the values of the hyperparameters, and therefore change whenever the hyperparameters are updated using Gibbs sampling, normally before each hybrid Monte Carlo update. In the standard hybrid Monte Carlo method, these updates begin with the complete replacement of the momentum variables, so the invariance of the distribution of $p_i/\sqrt{m_i}$ is of no significance. However, this is not the case for the variant of hybrid Monte Carlo with "persistence" discussed in Section 3.5.3.

A basis for choosing good stepsizes can be found by examining the behaviour of the leapfrog method applied to a simple system with a single position component (and hence a single momentum component) with the Hamiltonian $H(q,p) = q^2/2\sigma^2 + p^2/2$. A leapfrog step for this system is

$$\hat{p}(\tau + \tfrac{\epsilon}{2}) = \hat{p}(\tau) - \frac{\epsilon}{2} q(\tau)/\sigma^2 \tag{3.28}$$

$$\hat{q}(\tau + \epsilon) = \hat{q}(\tau) + \epsilon \hat{p}(\tau + \tfrac{\epsilon}{2}) \tag{3.29}$$

$$\hat{p}(\tau + \epsilon) = \hat{p}(\tau + \tfrac{\epsilon}{2}) - \frac{\epsilon}{2} q(\tau + \epsilon)/\sigma^2 \tag{3.30}$$

This defines a linear mapping from $(\hat{q}(\tau), \hat{p}(\tau))$ to $(\hat{q}(\tau + \epsilon), \hat{p}(\tau + \epsilon))$. By examining the properties of this mapping, it is straightforward to show that $H(\hat{q}, \hat{p})$ diverges if this leapfrog step is repeatedly applied with $\epsilon > 2\sigma$, but that H remains bounded when it is applied with $\epsilon < 2\sigma$. Setting ϵ somewhat below 2σ will therefore keep the error in H small, and the rejection rate low, regardless of how long the trajectory is.

This simple system serves as an approximate model for the behaviour of the leapfrog method when applied to a more complex system whose potential energy function can locally be approximated by a quadratic function of q. By a suitable translation and rotation of coordinates, such a quadratic energy function can be put in the form

$$E(q) = \sum_i q_i^2 / 2\sigma_i^2 \tag{3.31}$$

In this form, the components are independent under the canonical distribution, and do not affect one another in the leapfrog steps — the behaviour of each pair, (q_i, p_i), is as for the simple system considered above. However, the final decision to either accept or reject the result of following a trajectory is based on the total change in H, to which all components contribute.

Chapter 3. Monte Carlo Implementation

If we use the same stepsize for all components in this system, then to keep the rejection rate low, we will have to use a stepsize less than $2\sigma_{\min}$, where $\sigma_{\min}$ is the smallest of the σ_i, as otherwise the error in H due to one or more components will diverge as the trajectory length increases. If other of the σ_i are much larger than $\sigma_{\min}$, then with this small stepsize a large number of leapfrog steps will be required before these other components change significantly.

This inefficiency can be avoided by using a different stepsize for each component (equivalently, a different mass). For the ith component, we can set the stepsize, ϵ_i, to a value a bit less than $2\sigma_i$, with the result that even short trajectories traverse the full range of all components.

In practice, this result is too much to hope for, both because the potential energy is at best only approximately quadratic, and because we do not know how to translate and rotate the coordinate system so as to remove the interactions between components of q. Nevertheless, using a different stepsize for each component will generally be advantageous.

In this implementation, I use a heuristic approach in which the stepsizes are set as follows:

$$\epsilon_i \approx \eta \left[\frac{\partial^2 E}{\partial q_i^2} \right]^{-1/2} \qquad (3.32)$$

where η is a *stepsize adjustment factor*, chosen by the user. If the energy really were as in equation (3.31), the heuristic would give $\epsilon_i \approx \eta \sigma_i$, which is close to optimal when $\eta \approx 1$. When the different components interact, however, these stepsizes may be too large, and the user may need to use a smaller value for η in order to produce an acceptable rejection rate.

Unfortunately, we cannot set the stepsizes based on the actual values of $\partial^2 E / \partial q_i^2$ at the starting point of the trajectory. Doing this would render the method invalid, as the trajectory would cease to be reversible — when starting at the other end, different stepsizes would be chosen, leading to a different trajectory. We are allowed to use the current values of the hyperparameters, which are fixed during the hybrid Monte Carlo update, as well as the values of the inputs and targets in the training cases, but we must not use quantities that depend on the network parameters.

Details of the heuristic procedure for setting the ϵ_i using permissible information are given in Appendix A (Section A.4). The difficult part is the estimation of $-\partial^2 L / \partial w_{ij}^2$, where L is the log likelihood due to a training case, and w_{ij} is a weight in the network. Such estimates are obtained by first estimating $-\partial^2 L / \partial v_j^2$, where v_j is the value of a unit in the network. These estimates are found by a form of backpropagation, which need be done only once, not for every training case, since we are not permitted to use the actual values of v_j for a particular case anyway. Several heuristic approximations are made during this procedure: when a value depends on

v_j, the maximum is sometimes used; when the sign of a term depending on v_j may be either positive or negative, it is replaced by zero, on the assumption that these terms will ultimately cancel when we sum the results over the training set; and when a value depends on the magnitude of a weight, the magnitude corresponding to the associated hyperparameter is used. To find $-\partial^2 L/\partial w_{ij}^2$ based on $-\partial^2 L/\partial v_j^2$, we need the value of v_i^2. When v_i is an input unit, this value is available (since the inputs are fixed); when v_i is a hidden unit, the maximum possible value of $v_i^2 = 1$ is used.

3.2.3 Verifying correctness

The Markov chain Monte Carlo implementation described above is fairly complex, raising the question of how one can verify that the software implementing the method is correct.

One common type of implementation error results in answers that are correct, but require more computation time to obtain than they should have. In this respect, note that the validity of the hybrid Monte Carlo method requires only that the dynamics be reversible and preserve volume in phase space, and that the end-point of the trajectory be accepted or rejected based on a correct computation of the change in total energy. Errors computing the derivatives of E used in the leapfrog method do not invalidate the results, but will usually result in a large error in the trajectory and a consequent high rejection rate. (For severe errors, of course, the resulting inefficiencies may be so great that the Markov chain does not converge in any reasonable amount of time, and so no answers are obtained.)

Once a feel for correct behaviour is obtained, such errors can often be recognized by the anomalously high rejection rate, which can be reduced only by using a very small stepsize adjustment factor, or by using very short trajectories. The correctness of the derivative computation can then be tested by comparison with the results obtained using finite differences (a check commonly done by users of other neural network procedures as well). One can also look at the effect of reducing the stepsize while increasing the number of leapfrog steps to compensate; with a correct implementation the computed trajectory should reach a limit as the stepsize is reduced. This latter check may also reveal errors in the trajectory computation itself.

Incorrect answers may be produced as a result of errors in other components of the implementation, such as in the computation of the total energy used in deciding whether to reject, or in the Gibbs sampling updates for the hyperparameters. Such answers may sometimes be obviously ridiculous, but other times they may appear reasonable. To detect such errors, we need to compare with the answers produced using a method that is as

74   Chapter 3. Monte Carlo Implementation

far as possible independent of that being tested, and which preferably is simpler, and hence less likely to be erroneously implemented.

I have used the method of rejection sampling from the prior for this purpose. (This method was also used to produce the illustration in Section 1.2.4. Rejection sampling in general is discussed by Devroye (1986).) This method produces a sample of independent values from the posterior given the training data, from which Monte Carlo estimates can be computed, and compared with those obtained using the dependent values produced by a Markov chain method. These independent values from the posterior are obtained by generating independent values from the prior and then rejecting some of these with probability proportional to the likelihood due to the training data, with the scaling factor for the likelihood chosen so that the maximum possible rejection probability is one. (When a regression model is used in which the noise level is a hyperparameter, the likelihood has no upper bound, so the method must be modified slightly, as described in Appendix A, Section A.5.)

The rejection rate with this method can be extremely high. It can be feasibly applied only to very small training sets, with priors carefully chosen to give a high probability to parameter values that are well-matched to the data. For the test to be sensitive, large samples from the posterior must be obtained using both the rejection sampling method and the Markov chain Monte Carlo method being tested. I have performed these tests only for some simple network models with one hidden layer, which do not exercise all features of the implementation. Nevertheless, I expect that with a fair amount of effort it will usually be possible to use rejection sampling to test the correctness of the implementation when applied to a specific network model of interest for some application. Of course, subtle errors whose effects are fairly small may remain undetected, but these tests can provide some confidence that the results are not grossly in error.

3.3   A demonstration of the hybrid Monte Carlo implementation

To illustrate the use of the implementation based on hybrid Monte Carlo, and provide an idea of its performance, I will show here how it can be applied to learning a neural network model for the "robot arm" problem used by Mackay (1991, 1992b) to illustrate his implementation of Bayesian inference based on Gaussian approximations. This problem was also used in my tests of earlier hybrid Monte Carlo implementations (Neal 1992b, 1993a).

All timing figures given in this section are for an implementation written in C and run on an SGI Challenge D machine, with a MIPS R4400 CPU and

R4010 FPU, running at 150 MHz. The code was written with reasonable attention to efficiency, but was not fanatically tuned. Evaluation of the tanh activation function for hidden units was done using the standard library routine; use of fast approximations based on table lookup can lead to a build-up of error over long trajectories.

3.3.1 The robot arm problem

The task in the robot arm problem is to learn the mapping from joint angles to position for an imaginary "robot arm". There are two real input variables, x_1 and x_2, representing joint angles, and two real target values, y_1 and y_2, representing the resulting arm position in rectangular coordinates. The actual relationship between inputs and targets is as follows:

$$y_1 = 2.0 \cos(x_1) + 1.3 \cos(x_1 + x_2) + e_1 \qquad (3.33)$$

$$y_2 = 2.0 \sin(x_1) + 1.3 \sin(x_1 + x_2) + e_2 \qquad (3.34)$$

where e_1 and e_2 are independent Gaussian noise variables of standard deviation 0.05.

David MacKay kindly provided me with the training and test sets he used in his evaluations. Both these data sets contain 200 input-target pairs, which were randomly generated by picking x_1 uniformly from the ranges $[-1.932, -0.453]$ and $[+0.453, +1.932]$, and x_2 uniformly from the range $[0.534, 3.142]$.

The robot arm data is modeled using a network with one layer of tanh hidden units. The inputs connect to the hidden units, and the hidden units to the outputs; there are no direct connections from inputs to outputs. MacKay divides the parameters for this network into three classes — input-to-hidden weights, hidden unit biases, and hidden-to-output weights together with output unit biases — and uses three hyperparameters to control the standard deviations of Gaussian priors for parameters in each of these three classes. I used three analogous hyperparameters, but did not group the output unit biases with the hidden-to-output weights. Instead, I simply gave the output biases fixed Gaussian distributions with a standard deviation of one. This change in the model is motivated by the scaling properties discussed in Chapter 2, which show that while the magnitude of the hidden-output weights should go down as the number of hidden units increases, there is no reason for any corresponding change in the magnitude of the output biases.

In his work, MacKay gives the hyperparameters improper uniform distributions. This is not safe with a Markov chain Monte Carlo implementation, however, because the resulting posterior is also technically improper (though only because of its behaviour far from the region of high probability density). This is not a problem in MacKay's implementation, which sets the hyperparameters to single values, but would eventually result in divergent behaviour of a Markov chain sampling from the posterior.

Accordingly, I gave proper Gamma priors to the hyperparameters, represented by precision values, as in equation (3.17). In all three cases, the shape parameter used was $\alpha = 0.2$, which gives a fairly broad distribution, approximating the improper prior used by MacKay. The mean was $\omega = 1$ for the precision of input-to-hidden weights and hidden unit biases. For the precision of the hidden-to-output weights, I set ω to the number of hidden units, which is in accord with the scaling relationships discussed in Chapter 2.

I let the precision value for the noise (assumed the same for both targets) be a hyperparameter as well, with a Gamma prior as in equation (3.21), with $\alpha = 0.2$ and $\omega = 100$ (corresponding to $\sigma = 0.1$). MacKay fixes the noise level to the true value of $\sigma = 0.05$, but it seems more realistic to let the noise level be determined from the data.

3.3.2 Sampling using the hybrid Monte Carlo method

In this demonstration, Markov chain sampling from the posterior distribution was done using two phases, the first designed to reach a rough approximation to equilibrium quickly, the second to sample efficiently from that point. This is generally a good strategy, though the details of the two phases described below are not necessarily optimal for all problems.

In the *initial phase*, we start from some initial state, and simulate a Markov chain for as long as is needed for it to reach a rough approximation to the posterior distribution. In the *sampling phase*, we continue from the state reached at the end of the initial phase, generally using a different Markov chain, proceeding for long enough that a close approximation to the equilibrium distribution has been reached, and enough subsequent data has been collected to produce Monte Carlo estimates of adequate accuracy. Several runs of this two-phase procedure may be done, using different random number seeds; this provides a further check on whether equilibrium has actually been reached, as well as more data on which to base estimates.

In this section, I will demonstrate how these phases can be carried out for a network with 16 hidden units, applied to the robot arm problem with 200 training cases. The ultimate aim was to use the sample of networks obtained to make predictions for the targets in 200 test cases.

For both the initial phase and sampling phases, the Markov chain used was built by alternating Gibbs sampling updates for the hyperparameters (see Section 3.2.1) with hybrid Monte Carlo updates for the parameters (see Section 3.2.2). For the hybrid Monte Carlo updates, we must specify the number of leapfrog steps in a trajectory, L, and an adjustment factor,

η, for the heuristically chosen stepsizes. Typically, the best value for L is different for the initial phase and the sampling phase, which is one reason for having two phases.

Most of the computation time in this implementation goes to performing the leapfrog steps, since to evaluate the derivatives of E needed in each such step one must apply the network to all the training cases. Gibbs sampling for the hyperparameters and for the momentum variables takes comparatively little time. To facilitate comparison of runs in which the hybrid Monte Carlo trajectories consist of different numbers of leapfrog steps, I will present the results in terms of *super-transitions*, which may contain different numbers of hybrid Monte Carlo iterations, with different values of L, but which (for each phase) all contain the same number of leapfrog steps, and hence take approximately the same amount of computation time.[1] Within a super-transition, each hybrid Monte Carlo update is preceded by a Gibbs sampling update for the hyperparameters.

To investigate behaviour in the initial phase, I ran a series of tests using super-transitions in which a total of $2^{10} = 1024$ leapfrog steps were performed, in the form of 2^k hybrid Monte Carlo updates, each based on a trajectory of $L = 2^{10-k}$ leapfrog steps, with $0 \leq k \leq 10$. Each run started with the network parameters all set to zero; the initial values of the hyperparameters are irrelevant, since they are immediately replaced in the first Gibbs sampling update. I let each run go for twenty super-transitions, which took approximately 5.6 minutes of computation time in total. (Thus each leapfrog step took approximately 16 milliseconds.)

For all runs, the automatically assigned stepsizes were adjusted downwards by a factor of $\eta = 0.3$. A good value for η must be found by trial and error; as a rule of thumb, it should be set so that the rejection rate is roughly 20%. Alternatively, one might set η at random prior to each hybrid Monte Carlo update, using some moderately broad distribution.

Figure 3.2 shows the progress of these runs for $k = 0$, $k = 4$, and $k = 8$, which correspond to super-transitions consisting of a single hybrid Monte Carlo update with a trajectory of 1024 leapfrog steps, to 16 hybrid Monte Carlo updates with trajectories of 64 leapfrog steps, and to 256 hybrid Monte Carlo updates with trajectories of 4 leapfrog steps, together, in each case, with a like number of Gibbs sampling updates. Progress is shown in

---

[1] With the present implementation, this is not entirely true when L is very small, since the hyperparameters then change frequently, and whenever they do, the derivatives of E must be re-evaluated. This slowdown could be avoided by performing the Gibbs sampling updates less frequently, or by saving intermediate results that would allow the derivatives to be re-evaluated without re-examining all the training cases. Taking account of this slow-down for small L would in any case only strengthen the conclusions reached in this evaluation.

78   Chapter 3. Monte Carlo Implementation

Training error

[Figure: log-scale plot of training error vs. Number of super-transitions (5, 10, 15, 20), y-axis showing 1.0, 0.1, 0.01. Legend indicates three line styles for trajectories of 4, 64, and 1024 leapfrog steps.]

Number of super-transitions

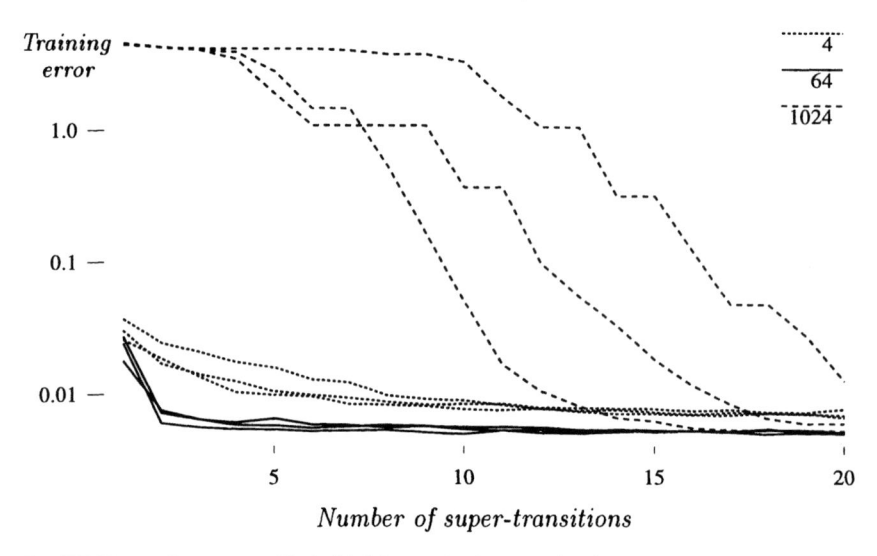

FIGURE 3.2. Progress of hybrid Monte Carlo runs in the initial phase. The plot shows the average squared error on the training set after each super-transition, on a log scale. The solid lines show the progress of three runs in which trajectories 64 leapfrog steps long were used. Dotted lines show the progress of three runs with trajectories of 4 leapfrog steps, dashed lines the progress of three runs with trajectories of 1024 leapfrog steps.

the figure in terms of the average squared error on the training set, which is closely related to the potential energy. The training set error was initially very high, since the network parameters had not adapted to the data. Once the training error had largely stabilized at a lower value, I assumed that the chain had reached at least a rough approximation to the equilibrium distribution, and that the sampling phase could begin.

As can be seen, convergence to a roughly equilibrium distribution was faster using trajectories consisting of 64 leapfrog steps than when using trajectories of 4 or 1024 leapfrog steps; trajectories of length 16 and length 256 were also inferior, though less dramatically so. This optimal trajectory length of 64 leapfrog steps is quite short in comparison with what will later be seen to be the optimum trajectory length for the sampling phase. This is understandable. The initial energy of the system is quite high, and must drop significantly for equilibrium to be reached. Energy is dissipated in the hybrid Monte Carlo method only when the momentum variables are replaced from their canonical distribution, which occurs only at the beginning of each hybrid Monte Carlo update, before the trajectory is computed. Rapid dissipation of energy therefore requires that many updates be done, with correspondingly short trajectories. The increased frequency of Gibbs sampling updates when trajectories are short may also contribute to faster

3.3 A demonstration of the hybrid Monte Carlo implementation 79

convergence. For very short trajectories, however, the slowing effect of the resulting random walk dominates.

Once the initial phase is complete, we can find good values for the step-size adjustment factor, η, and trajectory length, L, for use in the sampling phase. Prior to reaching a rough equilibrium at the end of the initial phase, it is possible that the situation will not have stabilized enough for this to be done.

Figure 3.3 shows data on how the error in total energy varies with η. This data was obtained by continuing the simulation from the state at the end of one of the initial phase runs, using values of η randomly selected from an interval of one order of magnitude around $\eta = 0.5$. Trajectories of length 100 were used here, but the results are similar for all but very short trajectories. As can be seen, for η greater than about 0.5, the leapfrog method becomes unstable, and very large (positive) errors result, which would lead to a very high rejection rate if such a value of η were used. The value $\eta = 0.3$ used in the initial phase is close to optimal and was therefore used for the sampling phase as well.

In order to minimize the extent to which the Markov chain undertakes a random walk, L should be chosen so that relevant functions of state at the end-point of a trajectory are almost uncorrelated with the corresponding values at the start-point. Trajectories should not be longer than is necessary to achieve this, of course.

Figure 3.4 shows the variation of several quantities along a single trajectory 10 000 leapfrog steps long, computed with $\eta = 0.3$, starting from the final state of one of the initial phase runs. The quantities directly relevant to the prediction task are the outputs of the network for the inputs in the test set; one such output is shown on the left of the figure. Though some short-range correlations are evident, these appear to die out within about 500 leapfrog steps, as is confirmed by numerical estimation, in so far as is possible from this small amount of data. A value of $L = 500$ might therefore seem appropriate for use in the sampling phase.

The right side of Figure 3.4 tells a different story, however. For each of the three classes of parameters for this network, it shows the variation along the trajectory of the square root of the average squared magnitude of parameters in that class. (These quantities determine the distribution of the hyperparameters associated with the classes.) Correlations are evident in these quantities over spans of several thousand leapfrog steps. Such long-term correlations are also found in the values of individual network parameters. These facts suggest that trajectories in the sampling phase should be several thousand leapfrog steps long (with $\eta = 0.3$).

One might question the need for such long trajectories, since the quantities exhibiting these long-range correlations are not of interest in them-

80   Chapter 3. Monte Carlo Implementation

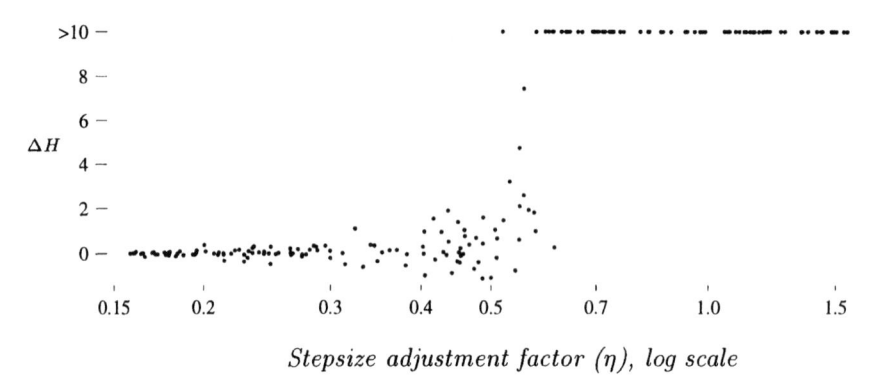

Stepsize adjustment factor (η), log scale

FIGURE 3.3. Error in energy for trajectories computed with different stepsizes. Each point plotted shows the change in total energy (H) for a trajectory of 100 leapfrog steps in which stepsizes were adjusted by the factor given on the horizontal axis (with changes greater than 10 plotted at 10). The starting point for the first trajectory was the last state from one of the initial phase runs with $L = 64$ shown in Figure 3.2. Starting points for subsequent trajectories were obtained by continuing the simulation using hybrid Monte Carlo with these trajectories, along with Gibbs sampling updates of the hyperparameters. Values for η were randomly generated from a log-uniform distribution.

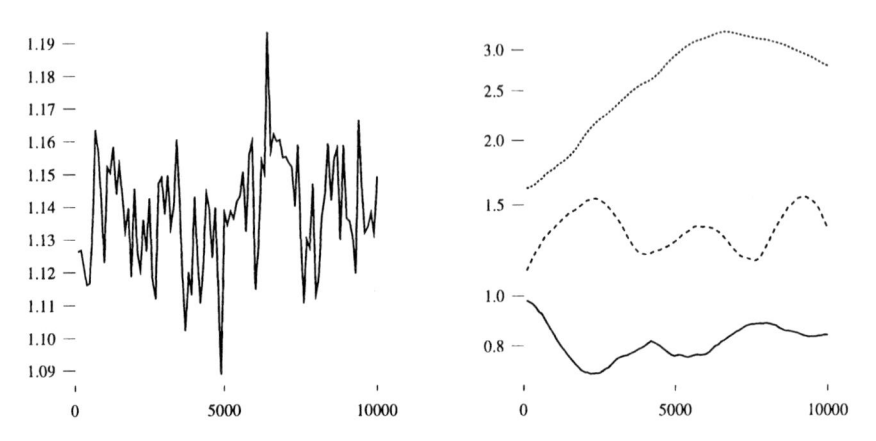

FIGURE 3.4. Degree of correlation along a trajectory. The plot on the left shows the first output of the network for inputs $(-1.47, 0.75)$, as the network parameters vary along a trajectory 10 000 leapfrog steps long (with $\eta = 0.3$), plotted every 100 steps. On the right, the variation in the square root of the average squared magnitude of parameters in three classes is shown — for input-hidden weights (solid), hidden biases (dotted), and hidden-output weights (dashed) — plotted on a log scale. The trajectory began with the state at the end of one of the initial phase runs with $L = 64$ shown in Figure 3.2.

3.3 A demonstration of the hybrid Monte Carlo implementation 81

selves. It is nevertheless prudent to pay attention to these quantities, for two reasons.

First, the initial phase produces a state that is only presumed to be from roughly the equilibrium distribution. Further exploration of the state space in the sampling phase may reveal that the true equilibrium distribution is in fact quite different; alternatively, if this does not happen, our confidence that the true equilibrium has been found is increased. For this purpose, the sampling phase should explore regions of state space that are representative of the posterior distribution in all relevant respects, which must certainly include aspects related to the hyperparameter values.

Second, even if autocorrelations for the quantities of interest appear from a short segment of the chain to go to zero fairly rapidly, as in the left of Figure 3.4, it is possible that if the values were examined over a longer period, significant long-term correlations would be evident. It is difficult to ever be completely sure that this is not the case, but here again confidence can be increased by ensuring that the chain explores the full range of hyperparameter values.

Figure 3.5 shows several sampling phase runs, with different trajectory lengths, each continuing from the state at the end of one of the initial phase runs with $L = 64$. For these runs, I used super-transitions consisting of 32 000 leapfrog steps. For the run using trajectories of length $L = 125$, each super-transition therefore consisted of 256 pairs of Gibbs sampling and hybrid Monte Carlo updates; for the run with $L = 2000$, each super-transition consisted of 16 pairs of updates; and for the run with $L = 32 000$, each consisted of a single Gibbs sampling update followed by a single hybrid Monte Carlo update. The state at the end of each super-transition was saved for possible later use in making predictions. The rejection rate for hybrid Monte Carlo updates was about 13% in all runs. Each run took approximately nineteen hours of computation time.

The results of these runs show that the initial phase had not fully converged to the equilibrium distribution. Equilibrium does appear to have been reached after about 50 sampling phase super-transitions for the run with $L = 125$, and after about 25 super-transitions for the runs with $L = 2000$ and $L = 32000$.

The run with $L = 2000$ clearly explored the range of these quantities more rapidly than did the run with $L = 125$. The relative merits of $L = 2000$ and $L = 32 000$ are less evident. To get a better idea of the effect of varying L, I did three independent sampling runs of 150 super-transitions with L set to each of 125, 500, 2000, 8000, and 32 000, in each case starting from the end states of the three initial phase runs with $L = 64$ shown in Figure 3.2. For each value of L, I used the data from the three runs to estimate the autocorrelations in the square root of the average squared

82   Chapter 3. Monte Carlo Implementation

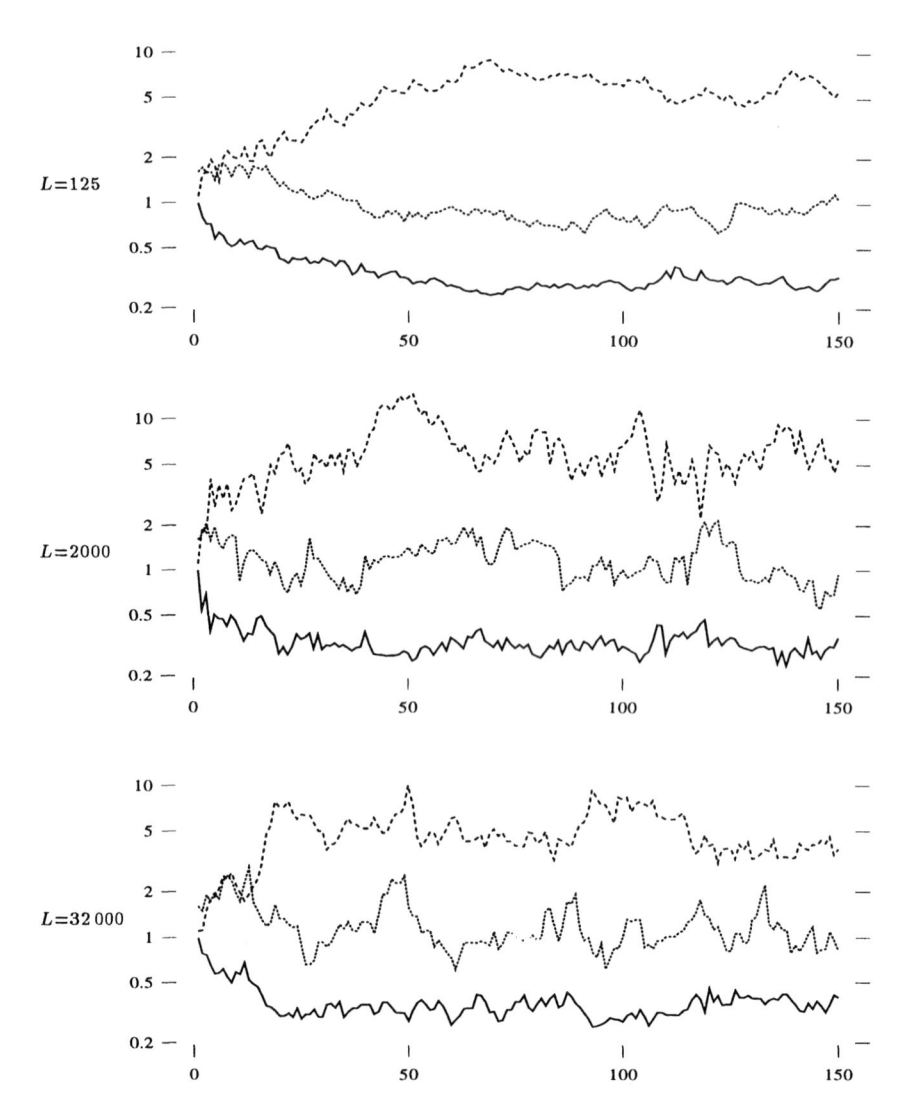

FIGURE 3.5. Progress of hybrid Monte Carlo runs in the sampling phase. These plots show the variation in the square root of the average squared magnitudes of parameters in the three classes during the course of hybrid Monte Carlo sampling runs using various trajectory lengths (L). The stepsize adjustment factor was $\eta = 0.3$ in all cases. The runs were started with the state at the end of one of the initial phase runs with $L = 64$ shown in Figure 3.2. The horizontal axes show the number of super-transitions, each consisting of 32 000 leapfrog steps. The vertical axes show the square roots of the average squared magnitudes on a log scale, with input-hidden weights shown with solid lines, hidden biases with dotted lines, and hidden-output weights with dashed lines.

3.3 A demonstration of the hybrid Monte Carlo implementation

| Input-Hidden Weights | Hidden Biases | Hidden-Output Weights |

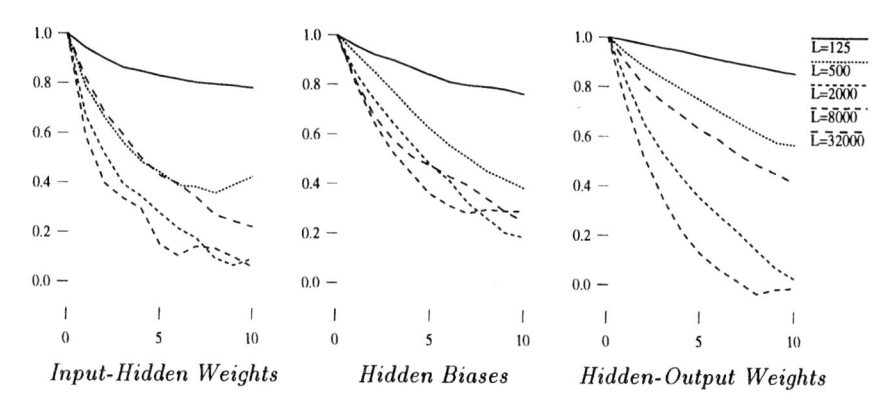

FIGURE 3.6. Autocorrelations for different trajectory lengths. The plots show autocorrelations for the square root of the average squared magnitude of network parameters in each of three classes. The horizontal axes give the lags, measured in super-transitions consisting of 32 000 leapfrog steps; the vertical axes show the estimated autocorrelations at these lags, for sampling runs that have reached equilibrium. Results are shown for runs in which the hybrid Monte Carlo updates use various trajectory lengths (L), as indicated.

magnitude of the parameters in different classes. In making these estimates, data from the first 50 super-transitions in each run was discarded, as the equilibrium distribution may not have been reached by then.

The results are shown in Figure 3.6. Trajectories of length $L = 8000$ have the smallest autocorrelations, though $L = 2000$ is not much worse. This is as anticipated from the trajectory plot in the right of Figure 3.4, showing that a reasonable value for L can be selected before extensive computations are done.

I have done some preliminary experiments to investigate why the autocorrelations for quantities shown in Figure 3.6 are non-zero (for lags greater than zero) even for the best sampling runs, with $L = 8000$. Three runs of 150 super-transitions with $L = 8000$ were done in which there was only a single Gibbs sampling update for the hyperparameters at the start of each super-transition. (Recall that in a normal sampling run with $L = 8000$, a Gibbs sampling update is done before each of the four hybrid Monte Carlo updates in a super-transition.) Autocorrelations for the square roots of the average squared magnitudes of input-hidden weights and hidden-output weights (but not hidden biases) were significantly greater in these runs than in the normal runs. The observed autocorrelations were in fact consistent with the hypothesis that these autocorrelations are determined entirely by the frequency of Gibbs sampling updates, as autocorrelations at lag 4ℓ in these runs were similar to autocorrelations at lag ℓ in the normal runs. In further sampling runs with a single Gibbs sampling update in each super-transition but with twice as many hybrid Monte Carlo updates (tak-

84   Chapter 3. Monte Carlo Implementation

ing twice as much time), the autocorrelations were reduced only slightly, adding further support to the hypothesis that the Gibbs sampling component of the Markov chain is the primary cause of the autocorrelations seen.

These results suggest that performance might be improved by merging the updates of the hyperparameters with the updates of the parameters. Such a scheme might be aimed at increasing the frequency of hyperparameter updates, or at suppressing the random walk nature of these updates by performing them using hybrid Monte Carlo. However, one would like to preserve the capability in the present implementation of using the hyperparameter values to set stepsizes for the parameter updates; this requirement makes devising such a scheme non-trivial.

3.3.3 Making predictions

Once we have one or more realizations of the Markov chain from the sampling phase, we can make predictions for test cases by using the states from these realizations as the basis for Monte Carlo estimates. States prior to when equilibrium was apparently reached should be discarded. Each state after equilibrium gives us values for the network parameters and hyperparameters that come from the posterior distribution given the training data (equation 3.13).

The sample from the posterior can be used directly to obtain a sample from the predictive distribution for the targets in a test case (equation 3.14), which may be useful in visualizing the predictive distribution, as well as being the basis for numerical estimates. The process is illustrated in Figure 3.7. We first compute the outputs of the network with the given test inputs for the values of the network parameters taken from the equilibrium portion of the sampling phase run (or runs). For the plot on the left of the figure, the last 100 states of one run were used, the first 50 being discarded in case they were not from the equilibrium distribution. In the model being used, the actual targets are obtained from these outputs by adding Gaussian noise, with a standard deviation (the same for both outputs) given by a hyperparameter that is also being estimated. To each of the 100 output values, we therefore add Gaussian noise with standard deviation given by the hyperparameter value that is part of the same state, to produce the sample from the predictive distribution shown on the right of the figure. (The posterior distribution of the noise standard deviation had a mean of 0.051, with a standard deviation of 0.002; recall that the true value used to generate the data was 0.05.)

If we need to make single-valued guesses for the targets in a test case, with the aim of minimizing expected squared error, we should guess the mean of the predictive distribution, which is the same as the mean value of

3.3 A demonstration of the hybrid Monte Carlo implementation

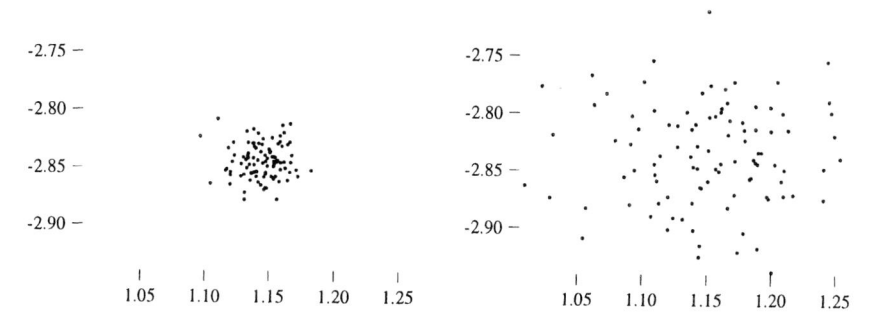

[scatter plots showing network output values, left plot clustered around (1.15, -2.85), right plot more dispersed, both with x-axis from 1.05 to 1.25 and y-axis from -2.90 to -2.75]

FIGURE 3.7. Predictive distribution from Monte Carlo data. The plot on the left shows the values of the two network outputs when the inputs are set to $(-1.471, 0.752)$ and the network parameters are taken from the last 100 super-transitions of one of the hybrid Monte Carlo sampling runs with $L = 8000$. The plot on the right shows the same 100 values with Gaussian noised added, with the standard deviation of the noise being determined in each case by the value of the noise-level hyperparameter at that point in the run; this plot represents the predictive distribution for the target values with these inputs. (The true relationship of equation (3.34) gives outputs (before noise) of $(1.177, -2.847)$ for these inputs.)

the network outputs. We can estimate this mean by simply averaging the network outputs for the values of the parameters taken from the sampling phase runs. The accuracy of such a Monte Carlo estimate is determined by the variance of the quantity whose mean is being estimated, the number of points in the sample, and the autocorrelations between these points, as is discussed in Section 1.3.1. In the example here, the autocorrelations of networks outputs for test cases from one super-transition to another in the sampling phase turn out to be quite small (assuming, as always, that there are no undetected long-range correlations). Accordingly, the variance of the estimate is just the variance of the output divided by the number of sample points, 100 here. For the test case illustrated in Figure 3.7, the estimated predictive means, with standard errors, are 1.1446 ± 0.0015 and -2.845 ± 0.0015. (Note that the accuracy of the Monte Carlo estimate of the predictive mean does not tell us what the likely error is when using this mean as a guess for the actual target values. The latter might be estimated by the standard deviation of the predictive distribution, but this estimate may be bad if the model is bad.)

The relationship between the predictions of the model and the actual targets in test cases is the subject of Chapter 4, but it is of interest here to compare the test error for the robot arm problem using the hybrid Monte Carlo implementation with the test error found by MacKay (1991, 1992b) using his implementation based on Gaussian approximations. (But note that the model I used is slightly different than that MacKay uses, as explained in Section 3.3.1.) Figure 3.8 shows the test error for the different

	Average squared test error		
Gaussian approximation method of MacKay			
Solution with highest evidence	0.00573		
Solution with lowest test error	0.00557		
Hybrid Monte Carlo, with 150 super-transitions			
Last 100 points from individual runs	0.00558	0.00554	0.00561
Last 100 points from all three runs		0.00557	
Hybrid Monte Carlo, with 30 super-transitions			
Last 15 points from individual runs	0.00557	0.00562	0.00560
Last 15 points from all three runs		0.00558	

FIGURE 3.8. Average test error on the robot arm problem with different implementations. The hybrid Monte Carlo sampling runs used super-transitions of 32 000 leapfrog steps each, with $L = 8000$ and $\eta = 0.3$.

implementations, measured as the average over the 200 test cases of the total squared error in guessing the two targets. The expected test error for guesses based on knowledge of the true distribution is 0.00500.

The test errors for MacKay's Gaussian approximation method are taken from a figure in his paper.[2] MacKay trains networks from many random starting points, finding many local minima, and evaluates the quality of each run by an "evidence" measure. In the top section of Figure 3.8, I give the test error both for the network of MacKay's with the largest evidence, and for the network with the smallest test error (but slightly lower evidence). The network with smallest test error cannot be identified from the training data, of course, but it is possible that a similarly small test error could be obtained by averaging the outputs of several of the networks with large evidence.

The middle section of Figure 3.8 shows results based on networks from the last 100 super-transitions of the hybrid Monte Carlo sampling runs described previously, with $L = 8000$. Results were very similar using the other runs with $500 \leq L \leq 32000$, but slightly worse for $L = 125$. The first results shown are for guesses found by averaging the outputs of the 100 networks in each run separately. There is little variation over the three runs, an indication that these runs had all reached a good approximation to the true equilibrium distribution and had sampled from its entirety. Since the guesses made here are based on Monte Carlo estimates of the predictive means, rather than the exact values implied by the model, the

---

[2]See Figure 11 of (MacKay 1992b). MacKay reports test performance in terms of the total squared error on the test set, scaled so that the expected total error based on the true relationship is equal to the total number of test targets. To convert his figures to average squared error, divide by 400 and multiply by 0.0050.

3.3 A demonstration of the hybrid Monte Carlo implementation

average squared error will be larger than that which would be obtained using an exact implementation — specifically, the expected squared error on a single test case will be inflated by the variance of the Monte Carlo estimate for the predictive mean for that case. The test error that results when the networks from all three runs are combined is shown in the figure as well; it differs little from the results of the separate runs. This provides additional evidence that equilibrium had been reached. It also shows that the inflation of the squared error due to the variance of the estimates is negligible in this example.

As can be seen, the test error using the hybrid Monte Carlo implementation is a bit better than for the network of MacKay's with the largest evidence (though no better than the network of MacKay's with lowest test error). It is tempting to regard this as an indication that the guesses found using hybrid Monte Carlo are closer to the true Bayesian predictions, though there is no theoretical guarantee that the true Bayesian predictions will have lower test error. The difference is rather small, however, so it appears that MacKay's Gaussian approximation was indeed adequate for the robot arm problem.

3.3.4 Computation time required

Solving the robot arm problem using one of these hybrid Monte Carlo runs requires nearly twenty hours of computation time — nineteen hours for the 150 super-transitions in the sampling phase, plus a bit for the initial phase and for chosing good values of L and η to use in the sampling phase. One may wonder whether this much computation time is really necessary to solve the problem using hybrid Monte Carlo. The bottom section of Figure 3.8 shows the test error obtained using the first 30 super-transitions of the sampling runs, with only the last 15 states of each run used in the estimates, earlier states being discarded in case they are not from the equilibrium distribution. As can be seen, the results from these shorter runs, each requiring about four hours of computation time, are not appreciably different from those based on the longer runs.

Unfortunately, it is only in retrospect that we can be sure that these short runs give good results. The first 30 super-transitions of the runs provide no clear evidence that equilibrium had been reached, though from the longer runs it appears that it had. Nevertheless, it may be necessary to use such short runs if more time is not available. Indeed, much more drastic abbreviations of the procedure can be contemplated. For example, averaging the outputs of the final five networks from all three initial phase runs with $L = 64$ shown in Figure 3.2 gives a test error of 0.00597. In some circumstances, this might be considered an acceptable result, obtained using about seventeen minutes of computation time.

It would be interesting to know how the computation time for Bayesian learning using hybrid Monte Carlo compares with that using the Gaussian approximation method. David MacKay (personal communication, 1994) has informed me that finding a solution for the robot arm problem using his program for Bayesian neural network learning requires about six minutes of computation time on a machine (a SPARC 10) of power roughly comparable to that of the machine used for my tests. Perhaps ten such runs would be needed to have some confidence that a good local minimum has been found, for a total computation time of about one hour. David MacKay feels that improvements to the program might significantly reduce this time. The hybrid Monte Carlo method may thus be somewhat slower than the Gaussian approximation method on the robot arm problem. These timing figures should not be taken too seriously, however, since they are heavily influenced by the details of the machines and programs used, and by the effort expended to ensure that the answer arrived at is as good as is possible.

3.4 Comparison of hybrid Monte Carlo with other methods

I claimed earlier that the hybrid Monte Carlo method is superior to simple forms of the Metropolis algorithm and to the Langevin method, due to its avoidance of random walk behaviour. In this section I will substantiate this claim with regard to the robot arm problem. I will also investigate whether uncorrected dynamical methods offer any advantage for this problem.

Comparisons of performance are easiest during the sampling phase, once the situation has stabilized. I will first look at how well various methods sample the square root of the average squared magnitude of the hidden-output weights, which determines the distribution of the associated hyperparameter. Recall that this was one of the quantities used to assess sampling performance in Section 3.3.2.

Figure 3.9 shows this quantity being sampled first by a simple form of the Metropolis algorithm with a Gaussian proposal distribution, second by the Langevin method (i.e. hybrid Monte Carlo with $L = 1$), and third by hybrid Monte Carlo with $L = 2000$ (which was seen in Section 3.3 to be close to the optimal trajectory length). The heuristic procedure for determining stepsizes described in Section 3.2.2 was used for all methods. For the simple Metropolis method, the "stepsize" for a parameter was used as the standard deviation for its Gaussian proposal distribution (the mean being the current value). The proposed changes for different parameters were independent. Note that the super-transitions used here consisted of only 2000 leapfrog steps or Metropolis updates, compared to 32 000 for the super-transitions in the sampling phase described in Section 3.3.2.

3.4 Comparison of hybrid Monte Carlo with other methods 89

Simple Metropolis *Langevin Method* *Vs. Hybrid Monte Carlo*

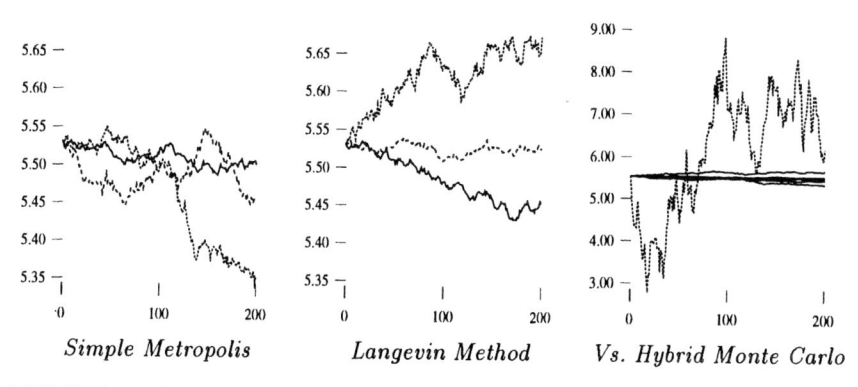

FIGURE 3.9. Progress of simple Metropolis and Langevin methods in the sampling phase. These plots show the square root of the average squared magnitude of the hidden-output weights for runs started from the equilibrium distribution (from the end of one of the sampling phase hybrid Monte Carlo runs). The horizontal axis gives the number of super-transitions, each consisting of 2000 simple Metropolis or Langevin updates for the parameters, or for the plot on the right, of a single hybrid Monte Carlo update using a trajectory 2000 leapfrog steps long. (In all cases, each parameter update is preceded by a Gibbs sampling update for the hyperparameters). On the left, results are shown for simple Metropolis with $\eta = 0.1$ (solid), $\eta = 0.3$ (dotted), and $\eta = 0.9$ (dashed). In the centre, results are shown for the Langevin method, with the same values of η. On the right, these results are re-plotted (solid) along with the result using hybrid Monte Carlo with $\eta = 0.3$ and $L = 2000$ (dotted). Note the change in vertical scale.

Results for the simple Metropolis method are shown in the left of Figure 3.9, with the stepsize adjustment factor, η, set to 0.1, 0.3, and 0.9. The acceptance rates with these values of η were 76%, 39%, and 4%, respectively. For $\eta = 2.7$, the acceptance rate was 0.04%, and performance was poorer than for any of the runs shown.

Results for the Langevin method are shown in the centre of Figure 3.9, again for η set to 0.1, 0.3, and 0.9. The acceptance rates were 99%, 81%, and 0.8%, respectively. No changes were accepted in a run with $\eta = 2.7$.

The plot on the right of Figure 3.9 shows that all these results are much worse than those obtained in a run using hybrid Monte Carlo with $\eta = 0.3$ and $L = 2000$. We can get a rough idea of how much worse the other methods are as follows. The width of the region explored by the simple Metropolis and Langevin runs in 200 super-transitions was in no case more than about 0.2. The hybrid Monte Carlo run explored a range of about 6, not much less than the full range seen in the longer runs of Figure 3.5. Since the simple Metropolis and Langevin runs operate via a random walk, for them to explore a similar range would likely require about $(6/0.2)^2 = 900$ times as many super-transitions as required for hybrid Monte Carlo.

90   Chapter 3. Monte Carlo Implementation

The difference in how well the methods sample was somewhat less dramatic for the quantities of direct interest, the outputs of the network for test cases, but it was still very substantial. As discussed in Section 1.3.1, the efficiency with which the expectation of a quantity can be estimated is determined by the sum of the autocorrelations for that quantity at all lags. For outputs in test cases, the sum of these autocorrelations was found to be a factor of ten or more greater for the simple Metropolis and Langevin methods than for hybrid Monte Carlo with $L = 2000$.

I have also tried using simple Metropolis and the Langevin method in the initial phase, with a variety of values for η. None of these runs came close to the performance of the hybrid Monte Carlo runs with $L = 64$ shown in Figure 3.2.

Might there be some way of getting simple Metropolis to perform better?

In an optimization context, Szu and Hartley (1987) advocate using a multivariate Cauchy rather than a Gaussian as the Metropolis proposal distribution. I have tried using a Cauchy proposal distribution for this problem and found the results to be little different from those described above using the Gaussian proposal distribution.

For many problems, the Metropolis algorithm can be made more efficient by using a proposal distribution in which only a small part of the state is altered. This is advantageous if the energy of the slightly altered state can be incrementally computed in much less time than would be required to find the energy of a completely new state. Such incremental computation is possible for neural networks with one output and one hidden layer; if appropriate intermediate results are saved, the outputs of such a network can be re-computed in constant time after a change in one weight. Optimistically, one might hope for an order of magnitude or more improvement in efficiency from using this technique in a simple Metropolis method. However, one could also try using this technique to speed up the computation of trajectories for hybrid Monte Carlo, so it is not clear that success here would change the relative merits of the two algorithms.

I have also investigated whether uncorrected stochastic dynamics (see Section 3.1.2) might have advantages over hybrid Monte Carlo. With hybrid Monte Carlo, the stepsize we can use is limited by the resulting rejection rate; for uncorrected stochastic dynamics, the stepsize is limited by our tolerance for the systematic error that inexact simulation introduces. In sufficiently large problems, we might expect that stochastic dynamics will have an advantage, since the error in the energy that controls the rejection rate will grow with system size, but the systematic error may perhaps not (for more on this, see the discussion by Toussaint (1989)). However, for the robot arm problem, I found that no significant benefit was obtainable using uncorrected stochastic dynamics, either with long trajectories, or with trajectories one step long (the uncorrected Langevin method). For stepsizes much larger than was used for the hybrid Monte Carlo runs, the trajectories became unstable, and the systematic error was very large. This is as one would expect from the data on the error in energy shown in Figure 3.3.

Uncorrected stochastic dynamics might still be of interest for reasons other than increased speed. Its greater simplicity might make it more attractive for hardware implementation, for instance. I have tried using uncorrected stochastic dynamics in a sampling phase run with $L = 8000$ and $\eta = 0.3$. This run was identical to the corresponding hybrid Monte Carlo run except that trajectories were never rejected. The results using uncorrected stochastic dynamics were essentially indistinguishable from those using hybrid Monte Carlo, showing that this is a viable option. I had previously obtained similar results with an earlier implementation (Neal 1993a). Nevertheless, I believe that hybrid Monte Carlo is the more robust choice for general use. When too large a stepsize is used with hybrid Monte Carlo, the result is the easily diagnosed problem of a high rejection rate; with uncorrected stochastic dynamics, the result is systematic error that might sometimes be significant, yet go unnoticed.

3.5 Variants of hybrid Monte Carlo

A number of variants of the hybrid Monte Carlo algorithm have been proposed. Some that might be useful in this application have not yet been evaluated, such as the use of discretizations of the dynamics other than the leapfrog method (Creutz and Gocksch 1989). I have made preliminary investigations into three variants — in which trajectories are computed using "partial gradients", in which a "windowed" acceptance procedure is used, and in which random walks are suppressed by using "persistence" rather by than by using long trajectories. These variations are not always better than the standard procedure, but they do give a significant advantage in some circumstances, especially when used together.

3.5.1 Computation of trajectories using partial gradients

When minimizing the training error for a neural network using gradient descent, many people do not compute the derivatives of the total error at each step, but instead look at only one training case, selected at random, or in sequence. (This is the method used in the original papers of Rumelhart, Hinton, and Williams (1986a, 1986b), for example.) In the limit of small stepsizes, this "on-line" learning procedure gives the same result as looking at all training cases each time, since at a small enough scale the error function will be close to linear, and the average effect of the on-line steps will be the same as that of a step based on the full training error. One might expect the on-line procedure to be superior when the training set is redundant, having many similar training cases.

A similar idea can be applied to the computation of trajectories for hybrid Monte Carlo. Taking a somewhat more general view, let us assume that we have K approximations to the potential energy function, given by $E_k(q)$, for $k = 1, \ldots, K$, and that the average of these approximations gives the true energy function, i.e. $E(q) = (1/K) \sum_{k=1}^{K} E_k(q)$. We can now consider replacing each of the leapfrog steps based on derivatives of E, done with a stepsize of ϵ, by a sequence of K leapfrog steps using in sequence the derivatives of each of the E_k, what I will call the "partial gradients", each done with a stepsize of ϵ/K. I will call such a sequence of leapfrog steps based on partial gradients a "multi-leap" with K steps; a multi-leap with one step is just an ordinary leapfrog step. In order to preserve reversibility, it is necessary to randomly decide for each trajectory whether to perform the multi-leaps by using the E_k in ascending order or in descending order. Alternatively, one can select a random permutation of the E_k for each trajectory, which also insures against the possibility that some particular ordering might be especially bad. This is the method I used in the experiments described below. (It would also be valid to choose a random permutation for each multi-leap within a trajectory, but this leads to much larger errors.)

In the limit of small ϵ, the procedure using partial gradients should produce the same trajectory as the standard procedure using full gradients. Of more interest is what happens for larger ϵ. If the E_k are in fact all identical to E, then the new procedure will be stable up to values of ϵ that are K times larger than those under which the standard procedure is stable. With a suitable choice of ϵ, each multi-leap will then move K times as far as a single standard leapfrog step could. Presumably the E_k are not quite identical to E, but if they are good approximations to it, we may expect that we will be able to use a value of ϵ that is at least somewhat greater than that usable with the standard procedure.

Of course, this procedure will be advantageous only if the partial gradients are cheaper to compute than the full gradient. When E represents the log of the posterior probability, cheaper approximations can be obtained by looking at only part of the training set. We can rewrite the energy function of equation (3.23) as follows (setting $F(\gamma)$ to zero for simplicity):

$$E(\theta) = -\log P(\theta \mid \gamma) - \sum_{c=1}^{n} \log P(y^{(c)} \mid x^{(c)}, \theta, \gamma) \qquad (3.35)$$

$$= \frac{1}{K} \sum_{k=1}^{K} \left[-\log P(\theta \mid \gamma) - K \sum_{c \in G_k} \log P(y^{(c)} \mid x^{(c)}, \theta, \gamma) \right] \qquad (3.36)$$

3.5 Variants of hybrid Monte Carlo

where the G_k form a partition of the n-case training set (preferably, as close to an equal partition as is possible). We can therefore use approximations defined as follows:

$$E_k(\theta) = -\log P(\theta \mid \gamma) - K \sum_{c \in G_k} \log P(y^{(c)} \mid x^{(c)}, \theta, \gamma) \quad (3.37)$$

We choose K to be greater than one, but still much less than n. The cost of performing a multi-leap in which the derivatives of E_k are computed for each k will then be only slightly greater than the cost of a single standard leapfrog step in which the derivatives of E are computed once.

In order for the procedure as a whole to leave the desired distribution exactly invariant, the end-point of a hybrid Monte Carlo trajectory computed using these approximations must be accepted or rejected based on the exact value of E. If the trajectories used are long, as will usually be necessary, this full evaluation of E will be a small part of the total cost of a hybrid Monte Carlo update.

I have investigated the effects of using partial gradients for the robot arm problem, with 200 training cases, by looking at the error in H over a trajectory of 200 multi-leaps for various values of the stepsize adjustment factor, η. The results are shown in Figure 3.10, for $K = 1$ (the standard method), $K = 4$ (G_k of size 50), and $K = 16$ (G_k of size 12 or 13). As can be seen, trajectories computed with $K = 4$ remain stable up to about $\eta = 1.0$, twice the limit for stability with the standard method. Little or no further improvement is seen with $K = 16$, however. For small values of η, the error in H with $K = 4$ is larger than for $K = 1$. For values of η between 0.5 and 1.0, the error with $K = 4$ is smaller, since the standard procedure is unstable, but the error is still large enough to produce a rather low acceptance rate.

Because of this, it is difficult to obtain much net benefit from using partial gradients for this problem. For example, I tried using $\eta = 0.6$ and $L = 4000$ with $K = 4$, which should produce trajectories similar to those produced by the standard procedure with $\eta = 0.3$ and $L = 8000$, but using about half the computation time. Due to the larger error in H, however, the acceptance rate for these trajectories was only about 50%, whereas for the standard procedure is was about 85%. Considering that there is a bit more computational overhead with $K = 4$ than with $K = 1$, the cost per accepted trajectory is about the same.

More empirical and theoretical work is needed to better understand the effect of using partial gradients. It seems possible that significant benefits might be obtained when the training set is larger than is the case in the robot arm problem, or when the model or prior are different. Fortunately, it turns out that a significant benefit can be obtained even for the robot arm

94   Chapter 3. Monte Carlo Implementation

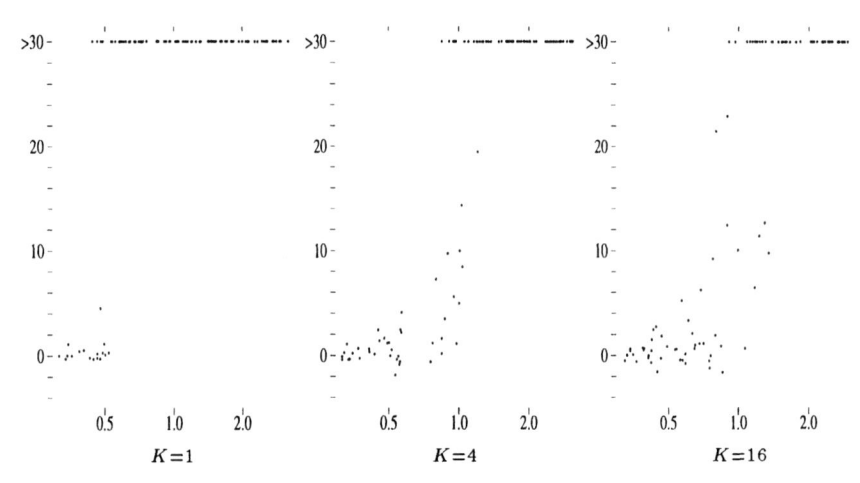

FIGURE 3.10. Error in energy for trajectories computed using partial gradients. Each plot shows the change in total energy (H) for 100 trajectories consisting of 200 multi-leaps with K steps. The plot on the left is for $K = 1$, the same as the standard method, for which data is also shown in Figure 3.3. The plots for $K = 4$ and $K = 16$ show the effect of using partial gradients. The horizontal axes show the randomly-selected stepsize adjustment factors (η) on a log scale; the vertical axes show the change in H, with changes greater than 30 plotted at 30. Starting points for the trajectories were obtained by using these trajectories in a hybrid Monte Carlo simulation, started at equilibrium.

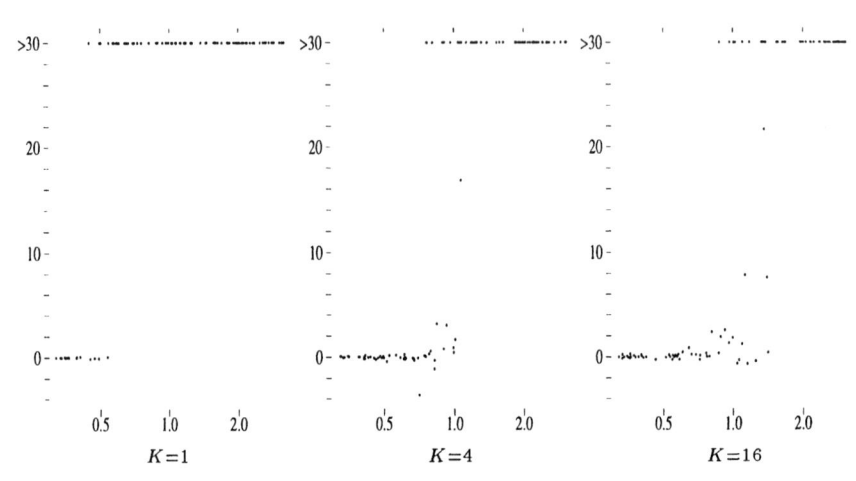

FIGURE 3.11. Difference in free energy for windowed trajectories. This figure is similar to Figure 3.10, but the trajectories were evaluated in terms of the difference in free energy between windows of length 20 at the beginning and end; this difference is shown on the vertical axes.

problem if the partial gradient method is combined with the "windowed" variant of hybrid Monte Carlo, which will be described next.

3.5.2 The windowed hybrid Monte Carlo algorithm

I have developed a variant of hybrid Monte Carlo in which transitions take place between "windows" of states at the beginning and end of a trajectory, rather than between single states (Neal 1994). Whether a candidate transition is accepted or rejected is based on the sum of the probabilities of the states in each window. This procedure has the effect of averaging over errors in H along the trajectory, increasing the acceptance rate. In this section, I will investigate the merits of this variant when applied to the robot arm problem, both when trajectories are computed by the standard method, and when they are computed using partial gradients.

In the windowed hybrid Monte Carlo algorithm, a trajectory computed by L leapfrog steps (or, if partial gradients are used, by L multi-leaps) is regarded as a sequence of $L+1$ states, in which the first W states constitute the "reject" window, $\mathcal{R}$, and the last W states the "accept" window, $\mathcal{A}$. The *free energy* of a window $\mathcal{W}$ is defined in analogy with statistical physics, as follows

$$F(\mathcal{W}) = -\log\left[\sum_{s \in \mathcal{W}} \exp\left(-H(q_s, p_s)\right)\right] \quad (3.38)$$

The sum of the probabilities of all states in a window is given, up to a constant factor, by $\exp(-F(\mathcal{W}))$, so the free energy plays the same role for windows as the total energy does for states.

Operation of the windowed algorithm is analogous to that of the standard algorithm — the momentum is randomized, a trajectory is computed, and the result of the trajectory is either accepted or rejected. In the windowed algorithm, however, the decision to accept or reject is based on the difference in free energies between the accept and reject windows. If the trajectory is accepted, the next state of the Markov chain is taken from the accept window, with a particular state from that window being selected at random according to their relative probabilities. Similarly, if the trajectory is rejected, the next state is randomly selected from the reject window.

It turns out that for this procedure to be valid, one further elaboration is required — the start state must be randomly positioned within the reject window. To accomplish this, we first choose an offset, T, for the start state uniformly from $\{0, \ldots, W-1\}$. We then compute the trajectory backwards from its normal direction for T leapfrog steps. (If the partial gradient method is used, we go backwards for T multi-leaps, during which the approximations must be applied in the reverse of their normal order.) Finally, after restoring the initial state, we compute the forward part of the trajectory, consisting of $L-T$ leapfrog steps (or multi-leaps).

The windowed algorithm can be used with a window size, W, up to the total number of states in the trajectory, $L + 1$. However, my tests on the robot arm problem were done only with windows much smaller than L; specifically, I used $W = 20$, while as seen in Section 3.3, the appropriate value of L is in the thousands. With such small windows, the distance moved when a trajectory is accepted is almost the same as for the standard algorithm with the same trajectory length. The two methods can therefore be compared by looking at their acceptance rates, which are determined by the differences in total energy or free energy between the start and end of the trajectory.

Figure 3.11 shows the difference in free energy between the accept and reject windows for 100 trajectories of length 200 started from the equilibrium distribution for the robot arm problem, for trajectories computed with full gradients ($K = 1$), and with partial gradients ($K = 4$ and $K = 16$). These plots correspond directly to those in Figure 3.10, done with the non-windowed algorithm. Comparing the two figures, it is clear that for trajectories that remain stable, the free energy differences for the windowed algorithm are significantly less than the total energy differences for the standard algorithm. As one would expect, there is no difference in the point at which the trajectories become unstable.

Accordingly, we should be able to use a larger value of η with the windowed algorithm than with the standard algorithm, while still maintaining a low rejection rate. For trajectories computed using the full gradient (on the left of the figures), this will give only a modest benefit, since the trajectories become unstable at about $\eta = 0.5$, not too far above the value $\eta = 0.3$ that was used in Section 3.3. (Note that in practice we would want to leave some safety margin between the value of η used and the point where the trajectories becomes unstable, since the point of instability will not be measured exactly and might vary during the course of the simulation.)

The windowed algorithm provides a significant benefit only when there is a significant range of stepsizes where the trajectories are not yet unstable, but do have large enough error that the acceptance rate is low. The size of this range should generally increase with the number of parameters (Neal 1994), so the windowed algorithm might be more useful with larger networks. The range of stepsizes giving stable trajectories with large error is also bigger when partial gradients are used, as seen in Figure 3.10. The centre and right plots of Figure 3.11 show that the windowed algorithm does indeed reduce the free energy differences in these cases.

To confirm that combining partial gradients with the windowed algorithm can give a significant benefit, I did three sampling phase runs with $K = 4$, $L = 4000$, $W = 20$, and $\eta = 0.6$, using super-transitions of 32 000 leapfrog steps, as in the Section 3.3. Trajectories of 4000 leapfrog steps with $\eta = 0.6$ should be equivalent to trajectories of 8000 leapfrog steps with $\eta = 0.3$, which were found in Section 3.3 to be of approximately optimal length.

Since twice as many trajectories are computed in a super-transition with $L = 4000$ than with $L = 8000$, we may hope for these runs to progress at twice the rate of the $L = 8000$ runs with the standard algorithm, as long as the rejection rate is not higher. The observed rejection rate using partial gradients and windows with $\eta = 0.6$ was approximately 15%, which is indeed close to the 13% rate seen for the standard algorithm with $\eta = 0.3$. As we would therefore expect, the runs using partial gradients and windows appeared to converge to the equilibrium distribution in about half the time (less than 10 super-transitions vs. around 15 or 20). Estimated autocorrelations for the quantities shown in Figure 3.6 were also as expected for a factor of two speedup.

3.5.3 Hybrid Monte Carlo with persistent momentum

I will briefly mention one further variation on hybrid Monte Carlo that I have used recently, due to Horowitz (1991), which I will refer to as hybrid Monte Carlo with "persistence" for the momentum.

Recall that the main advantage of hybrid Monte Carlo over other Markov chain methods is that random walks can be suppressed by using long trajectories, consisting of many leapfrog steps. In standard hybrid Monte Carlo, this advantage is lost if short trajectories are used, because the momentum variables are replaced in a Gibbs sampling step between each trajectory. Horowitz (1991) proposes using trajectories as short as a single leapfrog step, but with only partial replacement of the momentum variables between trajectories. Motion will therefore tend to "persist" in largely the same direction from one trajectory to the next, suppressing random walk behaviour.

An iteration of hybrid Monte Carlo with persistence operates as follows:

a) Perform a partial replacement of the momentum variables, setting them to new values, p'_i, as follows:

$$p'_i = \lambda p_i + (1 - \lambda^2)^{1/2} n_i$$

where n_i is a Gaussian random variate with mean zero and variance given by the mass, m_i, and λ is a parameter of the method, with a value between 0 and 1.

b) Perform a dynamical transition, as described by steps (1)–(3) on page 61 — briefly, one finds a candidate state by performing L leapfrog steps and then negating the momentum, and one then accepts or rejects this candidate state based on the change in H.

c) Negate the momentum variables, regardless of whether the candidate state was accepted in step (b).

Setting λ to zero in step (a) produces the equivalent of standard hybrid Monte Carlo. Setting λ to a value only slightly less than one produces a large degree of persistence, as the momentum variables will then be changed only slightly.

Steps (a) to (c) above will leave the canonical distribution invariant if each step does so individually. The dynamical transition of step (b) leaves the canonical distribution invariant for the reasons discussed previously. That steps (a) and (c) also leave the canonical distribution invariant can be seen directly from equation (3.2).

Step (c) of the procedure is crucial. Without it, the negation in step (b) will result in the path of each accepted trajectory being almost retraced when the next trajectory is accepted — not at all what we hope for from a method that is supposed to move persistently in one direction. The two negations cancel when the candidate state in step (b) is accepted. One might think that the method could be simplified by removing the negations from both step (b) and step (c), but the resulting method would not leave the canonical distribution invariant.

Note, however, that this method will perform well only if the rejection rate in step (b) is small, since when a rejection occurs, the state is left unchanged by step (b), but the momentum is still negated in step (c), causing an undesirable reversal of direction. Horowitz (1991) uses trajectories consisting of a single leapfrog step, which for a given stepsize produce a lower rejection rate than longer trajectories. I generally use trajectories of moderate length with the "windowed" acceptance procedure of Section 3.5.2; this can also produce a very low rejection rate.

Is there any advantage to be gained by suppressing random walks in this way rather than by using long trajectories? In a general context, it is not clear that there is any advantage. With this implementation of Bayesian neural network learning, however, reducing the length of trajectories allows Gibbs sampling updates for the hyperparameters to occur more often. As discussed at the end of Section 3.3.2, the frequency of hyperparameter updates may well be the limiting factor in this implementation.

My preliminary experience is that using shorter trajectories with persistence can indeed speed up sampling. Rasmussen (1996) has also obtained good results with this method. More systematic investigation of the merits of this and other variants of hybrid Monte Carlo is required, however.

Chapter 4
Evaluation of Neural Network Models

This chapter reports empirical evaluations of the predictive performance of Bayesian neural network models applied to several synthetic and real data sets. Good results were obtained when large networks with appropriate priors were used on small data sets for a synthetic regression problem, confirming expectations based on properties of the associated priors over functions. The Automatic Relevance Determination model was effective in suppressing irrelevant inputs in tests on synthetic regression and classification problems. Tests on two real data sets showed that Bayesian neural network models, implemented using hybrid Monte Carlo, can produce good results when applied to realistic problems of moderate size.

From a doctrinaire Bayesian viewpoint, a learning procedure is correct if it accurately captures our prior beliefs, and then updates these beliefs to take proper account of the data. If these prior beliefs are uninformative, or are actually wrong, the Bayesian procedure may have poor predictive performance, but the fault in such a case lies not with the procedure employed, but with our own ignorance or error. There might therefore seem to be no point in empirically testing Bayesian learning procedures; we should simply select a procedure that implements a model and prior that accord with our beliefs, as determined by careful introspection.

Whatever its merits in simple situations, this approach is clearly inadequate when using complex models such as neural networks. Although we can gain some insight into the nature of these models by theoretical analysis and by sampling from the prior, as was done in Chapter 2, we

100   Chapter 4. Evaluation of Neural Network Models

will probably never have a complete, intuitive understanding of their nature, and hence will never be entirely confident that our selection of such a model truly captures our prior beliefs. Furthermore, even complex models are seldom complex enough. We usually try to make do with a model that ignores certain aspects of our beliefs that we hope are not crucial for the problem at hand. This hope will not always be fulfilled.

Empirical testing therefore does have a role to play in the development of complex Bayesian models. It may reveal characteristics of the models that were not apparent to us initially, as well as identifying as crucial some aspects of the problem that we had at first hoped we could ignore. Testing is also needed in order to judge whether the implementation used is adequate. Finally, empirical performance is the only common ground on which Bayesian methods can be compared with those having a different philosophical basis.

In this chapter, I first use two synthetic data sets to evaluate a number of Bayesian neural networks models, using the Markov chain Monte Carlo implementation described in Chapter 3. One objective of these tests is to confirm that large networks perform well even with small training sets, as expected from the analysis in Chapter 2. Another aim is to investigate the performance of hierarchical models, particularly the Automatic Relevance Determination (ARD) model.

I then apply the Bayesian method to two real data sets, using models and priors selected in light of the discussions in Chapters 1 and 2, as well as the previous experience with synthetic data sets. These real data sets have been previously used in evaluations of other learning procedures, allowing some comparisons to be made between these procedures and the Bayesian models.

4.1   Network architectures, priors, and training procedures

The tests reported in this chapter used the network architectures and priors discussed in Chapters 1 and 2 and the hybrid Monte Carlo implementation of Chapter 3. I will briefly review these here. Additional details are also found in Appendix A.

The networks used are multilayer perceptrons with zero or more layers of tanh hidden units. The first hidden layer is connected to the inputs; subsequent hidden layers are connected to the previous hidden layer, and optionally to the inputs as well. The linear output units have connections from the last hidden layer (if present), and may also have direct connections from the input units. There are also biases for the hidden and output units.

4.1 Network architectures, priors, and training procedures

The outputs of the network define the conditional distribution of the target values associated with the inputs, according to whatever data model is being used.

The priors for the network parameters (the weights and biases) are defined hierarchically, using hyperparameters that control the standard deviations for weights and biases in various groups. In some cases, a single hyperparameter controls all the weights on connections from units in one layer to units in another layer (e.g. all weights from the input units to units in the first hidden layer). In other models, a finer level of control is exercised, with a separate hyperparameter being used to control the weights out of each unit of some type (e.g. all weights from one input unit to units in the first hidden layer).

In detail, suppose that $u_1, \ldots, u_k$ are the parameters (weights or biases) in one group. The hyperparameter associated with this group gives the standard deviation, σ_u, of a Gaussian prior for these weights:

$$P(u_1, \ldots, u_k \mid \sigma_u) \;=\; (2\pi)^{-k/2} \sigma_u^{-k} \exp\left(-\sum_i u_i^2 / 2\sigma_u^2\right) \quad (4.1)$$

The prior for the hyperparameter itself is expressed in terms of the "precision", $\tau_u = \sigma_u^{-2}$, which is given a prior distribution of the Gamma form, with mean ω_u:

$$P(\tau_u) \;=\; \frac{(\alpha_u/2\omega_u)^{\alpha_u/2}}{\Gamma(\alpha_u/2)} \tau_u^{\alpha_u/2 - 1} \exp\left(-\tau_u \alpha_u / 2\omega_u\right) \quad (4.2)$$

The value of α_u (which must be positive) controls how broad the prior for τ_u is, with the prior being broader for values of α_u close to zero. Note that the prior for $\sigma_u^2 = 1/\tau_u$ implied by equation (4.2) has a heavier upward tail than the prior for τ_u itself. Put another way, the prior for $\log \sigma_u$ has a heavier upward tail than downward tail. This asymmetry is probably undesirable; the Gamma form was chosen despite this because of its mathematical convenience.

Integrating over τ_u reveals that the prior for $u_1, \ldots, u_k$ is in fact a multivariate t-distribution, with α_u as its shape parameter. This way of viewing the prior is not particular useful when the parameter group consists of all the weights between units in two layers, but it can be when the prior is for a more specific group of weights. When the weights on the connections out of each hidden unit are treated as a separate group, with each hidden unit having an associated precision hyperparameter, the resulting t-distributions (with $\alpha_u < 2$) produce priors that, when properly scaled, converge to non-Gaussian stable distributions, and can thus be used in indefinitely large networks, as discussed in Chapter 2, and below in Section 4.2.[1] Using separate hyperparameters for the weights out of each input unit gives the Automatic Relevance Determination (ARD) prior discussed in Section 1.2.3, and below in Section 4.3.[2] In these cases, it is often desirable to add another level to the hierarchy by letting the mean precision for the weights (ω in equation 4.2) be a hyperparameter as well, common to a number of parameter groups of one type. This higher-level hyperparameter can then be given a prior of the same Gamma form.

Gibbs sampling and hybrid Monte Carlo were used to sample from the posterior distribution for the network parameters and hyperparameters, conditional on the training data, in the manner demonstrated in Section 3.3. Each run consisted of a short initial phase, whose purpose was to reach a rough approximation of equilibrium, and a much longer sampling phase, whose purpose was to reach a close approximation of equilibrium, and then to collect a sample of values from the posterior distribution of network parameters sufficient for making predictions. The sampling phases consisted of some number of "super-transitions", each of which consisted of some number of pairs of Gibbs sampling updates for the hyperparameters and hybrid Monte Carlo updates for the parameters. Only the states at the ends of the super-transitions were saved for possible later use in making predictions. The hybrid Monte Carlo trajectory length (L) and stepsize adjustment factor (η) were set differently for the two phases, based on trial and error and on tests following the initial phase. The "partial gradient" and "windowed" variants of hybrid Monte Carlo (see Section 3.5) were used for some problems. When partial gradients are used, I will use the phrase "leapfrog step" to refer to what was called a "multi-leap" in Chapter 3 — that is, a series of leapfrog steps that together look once at each training case.

Timing figures given in this chapter are for the same machine as was used for the demonstration in Section 3.3.

4.2 Tests of the behaviour of large networks

In Chapters 1 and 2, I argued that when using a properly-specified prior there is no need to limit the complexity of neural network models — indeed, in most circumstances, only an infinite network is truly capable of capturing

---

[1] The implementation also supports direct specification of t-distributions for individual parameters, but the indirect form may be preferable because τ_u can then be used in the heuristic procedure for setting stepsizes (see Section 3.2.2 and Section A.4).

[2] In an ARD network where inputs connect both to a hidden layer and directly to the outputs, each input unit will have two hyperparameters, controlling weights on connections to the two different layers. It might be desirable to link these two hyperparameters in some way, but the implementation does not support this at present.

our beliefs about the problem. In particular, I demonstrated in Chapter 2 that the prior over functions implied by a properly-scaled prior over weights will reach a limit as the number of hidden units in the network increases.

We would like to know more than was established theoretically, however. How many hidden units does it take to approach the limiting prior over functions? Is the limiting prior better for typical problems than a prior obtained using a small network? How well can the Markov chain Monte Carlo implementation handle large networks? Empirical testing can help in answering these questions.

4.2.1 *Theoretical expectations concerning large networks*

Before presenting empirical results using large networks, I will discuss the implications and limitations of the theoretical results of Chapter 2, in order to clarify what we might expect to see in the empirical tests.

First, note that though I advocate using networks with large number of hidden units (to the extent that this is computationally feasible), the arguments I present in Chapter 2 do not guarantee that increasing the number of hidden units in a network will always lead to results that are better than (or even as good as) those obtained with a small number of hidden units. No such guarantee is possible. If the function being learned happens to be tanh, for example, a network with one tanh hidden unit will perform substantially better than any more complex network. Even if the true function can only be exactly represented by an infinite network, it is possible that it is very close to a function that can be represented by a small network, in which case the small network may give better predictions when the training set is small, unless the prior used for the large network puts extra weight on those regions of the parameter space that produce functions close to those representable by a small network.

The theoretical arguments do show that large networks should behave "reasonably". By this I mean that they will neither grossly "overfit" the data — reproducing the targets in the training set very closely but performing poorly on test data — nor grossly "underfit" the data — ignoring the training set entirely. In empirical tests, we should therefore expect that performance using any of the properly-scaled priors discussed in Chapter 2 will reach a limit as network size increases, and in this limit performance will be reasonably good.

Many models will avoid the extremes of overfitting and underfitting, however, of which some will perform better than others. Sometimes a simple model may outperform a more complex model, at least when the training data is limited. Nevertheless, I believe that deliberately limiting the complexity of the model is not fruitful when the problem is evidently complex. Instead, if a simple model is found that outperforms some particular complex model, the appropriate response is to define a different complex model that captures whatever aspect of the problem led to the simple model performing well.

For example, suppose that on some problem a network with a small number of hidden units outperforms one with a large number of hidden units, using a Gaussian prior for the hidden-to-output weights. As seen in Chapter 2, a Gaussian prior for hidden-to-output weights leads to functions that are built up of contributions from many hidden units, with each individual hidden unit's contribution being insignificant. If a small network performs better than a large network when using this Gaussian prior, one may suspect that the prior is not appropriate. One might then hope that a large network using a prior based on a non-Gaussian stable distribution would better capture the properties of the problem, as it would allow a small number hidden units to have a large effect (as in a small network), while also allowing small corrections to these main effects to be made using additional hidden units.

4.2.2 Tests of large networks on the robot arm problem

I have tested the behaviour of Bayesian learning with large networks on the robot arm problem of MacKay (1991, 1992b), a regression problem with two input variables and two target variables, described in Section 3.3.1. For these experiments, I divided the 200-case training set used by MacKay into two training sets of 50 cases and one of 100 cases. Using these smaller training sets should make it easier to "overfit" the data, if overfitting is in fact a problem.

To evaluate predictive performance, I used a test set of 10 000 cases, drawn from the same distribution as the training data. Two performance criteria were used. First, following MacKay, I looked at the average over the test set of the sum of the squared errors for the two targets, when guessing the mean of the predictive distribution. Second, I looked at the average sum of the absolute errors for the two targets, when guessing the median of the predictive distribution. The second criterion is less sensitive to large errors. Since the targets are generated with Gaussian noise of standard deviation 0.05, the expected squared error on a single test case when using the optimal procedure based on the true relationship is $2 \times (0.05)^2 = 0.0050$.[3] The expected sum of absolute errors using the optimal procedure is $2 \times 0.80 \times 0.05 = 0.080$, where 0.80 is the expected absolute value of a variable with a standard Gaussian distribution.

---

[3] MacKay reports test performance in terms of the total squared error on a test set with 200 cases, scaled so that the expected total error based on the true relationship is equal to the total number of test targets. To convert his figures to average squared error, divide by 400 and multiply by 0.0050.

I modeled the robot arm data using networks with 6, 8, 16, and 32 tanh hidden units. (Preliminary experiments with networks containing only four hidden units showed that their performance was much worse.) Gaussian priors were used for the input-to-hidden weights and for the hidden biases; both Gaussian and Cauchy priors were tried for the hidden-to-output weights. The width parameters for these priors were controlled by hyperparameters, so that their values could adapt to the data, as would normally be desirable for real problems. The priors for the hyperparameters controlling the input-to-hidden weights and the hidden biases were the same for all networks; the prior for the hyperparameter controlling the hidden-to-output weights was scaled depending on the number of hidden units, in accord with the results of Chapter 2. For all three hyperparameters, the priors chosen were intended to be "vague". Improper priors were avoided, however, since they may lead to posterior distributions that are also improper. Very vague proper priors were avoided as well, partly because at some extreme a vague proper prior will suffer from the problems of an improper prior, and partly because of the possibility that with a very vague prior the Markov chain Monte Carlo implementation might become stuck for an extended period in some ridiculous region of the parameter space.

In detail, the precision (inverse variance) for the input-to-hidden weights was in all cases given a Gamma prior with mean precision of $\omega = 100$ (corresponding to a standard deviation of 0.1) and shape parameter $\alpha = 0.1$ (see equation 4.2).[4] The same prior was given to the precision hyperparameter for the hidden biases. The output biases were given a fixed Gaussian prior with standard deviation one. The prior for the hidden-to-output weights varied. When a Gaussian prior was used for hidden-to-output weights, the precision of the Gaussian was given a Gamma prior with $\alpha = 0.1$ and with mean $\omega = 100H$, where H is the number of hidden units (corresponding to scaling the standard deviation by $H^{-1/2}$). To implement a Cauchy prior for hidden-to-output weights, a 2-level scheme was used, as described in Section 4.1. For the low level, $\alpha = 1$ was used, to give a bivariate Cauchy distribution for the two weights out of each hidden unit.[5] For the high-level precision, used as the mean for the low-level precisions, a Gamma distribution with $\alpha = 0.1$ and with mean $\omega = 100H^2$ was used (corresponding to scaling the width of the Cauchy distribution by H^{-1}).

---

[4] In Chapter 3, I used priors with $\omega = 1$ and $\alpha = 0.2$. This turns out to be not as vague as is desirable, particularly in the direction of low variance. This is not crucial with 200 training cases (as in Chapter 3), but has a noticeable effect with only 50 training cases.

[5] One might instead give the two weights out of each hidden unit independent Cauchy distributions. In the limit of many hidden units, the two targets would then be modeled independently (see Section 2.2.1), except for the interactions introduced by the common hyperparameters. This model might well be better for this data, but it was not tried in these tests.

The noise level was the same for both outputs. It was controlled by a precision hyperparameter that was given a Gamma distribution with mean $\omega = 100$ and shape parameter $\alpha = 0.1$.

Learning began with a short initial phase, followed by a long sampling phase, as discussed in Section 4.1. The sampling-phase super-transitions consisted of ten pairs of Gibbs sampling and hybrid Monte Carlo updates. I used the partial gradient method (Section 3.5.1) for computing the hybrid Monte Carlo trajectories, with a five-way division of the training data, and the windowed acceptance procedure (Section 3.5.2), with a window size of ten. Stepsize adjustment factors were chosen so as to keep the rejection rate low (between 5% and 15%). Trajectory lengths were chosen to match the periods over which quantities such as the sum of the squares of the weights in various groups appeared to vary, in tests done following a few of the initial phase runs. The resulting choices were a stepsize adjustment factor of $\eta = 0.5$ and a trajectory of $L = 4000$ leapfrog steps for networks with 6, 8, and 16 hidden units, and $\eta = 0.4$ and $L = 5000$ for networks with 32 hidden units.

The number of sampling phase super-transitions needed to reach a good approximation to equilibrium was judged subjectively, largely by looking at the behaviour of the hyperparameters and of the squared error on the training set. On this basis, equilibrium may well have been reached in most cases after about 10 super-transitions, but I conservatively discarded the first 100 super-transitions for the networks with 8, and 16 hidden units, and the first 200 super-transitions for the networks with 6 and 32 hidden units. The smallest networks may require longer to reach equilibrium because the roles of the hidden units become constrainted, inhibiting movement about the parameter space; the largest networks may require longer because the larger number of parameters makes the Gibbs sampling updates of the hyperparameters less efficient.

For each network, I continued the sampling phase for an additional 200 super-transitions beyond the point where equilibrium was judged to have been reached. The 200 networks saved after these super-transitions were applied to each of the test cases, and the outputs used to make predictions. When guessing so as to minimize expected squared error loss, I averaged the outputs of the 200 networks, in order to estimate the mean of the predictive distribution for the targets in the test case. When guessing so as to minimize expected absolute error loss, I randomly generated five values from the target distribution defined by each network (a Gaussian with mean given by the network outputs, and standard deviation given by the current noise level), and then found the median of the resulting 5×200 target values, in order to estimate the median of the predictive distribution.

The accuracy of such estimates for the predictive means and medians depends not only on the sample size of 200, but also on the autocorrelations

Hidden units	Trajectory L	η	Super-transitions discarded	total	Time (hours) 50 cases	100 cases
6	4000	0.5	200	400	9	15
8	4000	0.5	100	300	8	14
16	4000	0.5	100	300	14	26
32	5000	0.4	200	400	46	81

FIGURE 4.1. Computational details for experiments on networks of varying size. The trajectory parameters shown are the number of leapfrog steps in a trajectory (L) and the stepsize adjustment factor (η). Also shown are the number of super-transitions discarded in order to reach equilibrium and the total number of super-transitions. These implementation choices varied with the number of hidden units, but not with the prior or with the number of training cases. The total computation time for all super-transitions is also shown; it does vary with the number of training cases.

of the network outputs for the test cases (see Section 1.3.1). For all combinations of network size and prior these autocorrelations were too small to reliably distinguish from zero on the basis of the data. Non-zero autocorrelations were observed for the hyperparameters, however, especially in the largest and smallest networks. For example, in the networks with 32 hidden units, the hyperparameter controlling the magnitude of input-to-hidden weights had substantial autocorrelations up to a lag of around five or ten super-transitions. Individual network parameters had substantial autocorrelations for the networks with 6 and 8 hidden units, but not for larger networks. These autocorrelations might lead one to suspect that there could be undetected autocorrelations for the output values as well, but these are presumably rather small. On this assumption, the sample of 200 networks is large enough that the degradation in performance due to the variance in the estimates of the predictive mean and median should be negligible; this is confirmed by the fact that the error when using only 100 of these networks is quite similar.

The computational details of the Markov chain Monte Carlo runs are summarized in Figure 4.1, which also gives the time required for these computations.

The predictive performance of Bayesian learning using the three training sets is shown in Figure 4.2, for networks with varying numbers of hidden units, using both Gaussian and Cauchy priors for the hidden-to-output weights. In all contexts, the networks with only 6 hidden units performed worse than the others, but no clear pattern of variation with network size can be seen amongst networks with 8 or more hidden units. On training set A, the networks with 8 hidden units perform better than those with 16 or 32 hidden units, but on training set B, of the same size, the reverse is true,

108   Chapter 4. Evaluation of Neural Network Models

showing that these differences are within the variation due to the random selection of training cases.

There is thus no reason to suspect that the larger networks were either "overfitting" or "underfitting" the data. Instead, as expected, performance with each training set appears to reach a reasonable limiting value as the size of the network increases. Lack of overfitting is also indicated by the estimates produced for the standard deviation of the noise in the targets. In all cases, the noise estimates were close to the true value of 0.05 — slightly higher than the true value for the small networks, quite close to the true value for the larger networks. If the larger networks were overfitting, one would expect their noise estimates to be substantially below the true value.

These results differ from those reported by MacKay (1991, 1992b), who found a slight decline in the "evidence" for larger networks (up to twenty hidden units) applied to the robot arm problem with a training set of 200 cases. (He also found that the evidence was correlated with performance on test data.) Although MacKay did not explicitly scale the prior for hidden-to-output weights as required for a limit to be reached as the number of hidden units increases, he did treat the variance for these weights as a hyperparameter. The variance should therefore have adopted the proper scaling automatically, allowing the large networks to perform well.

There are several possible explanations for this discrepancy. It is possible that the decline seen by MacKay was not indicative of a general and continuing trend — it might not have continued for still larger networks, and it might not have been seen on another training set. As I have noted, there is no guarantee that small networks will always perform worse than large networks; the reverse is seen in Figure 4.2 with training set A, though not with training set B. It is also possible that the Gaussian approximation method used by MacKay became inaccurate for the larger networks; I argued in Section 1.2.5 that this is to be expected.

Though Figure 4.2 shows no consistent differences in average squared error or average absolute error between networks with only 8 hidden units and those with 16 or 32 hidden units, a difference was apparent in the predictive distributions produced. As illustrated in Figure 4.3, predictions for test cases where the inputs were not close to those in any training case were consistently more uncertain in the larger networks — that is, the variance of the outputs of the network, plus the noise variance, was larger for the larger networks.[6] This is not unexpected. Since a small network will be able to represent only a limited range of functions, it will generally produce

---

[6] For test cases near to cases in the training set, the variance of the network outputs was also generally larger for the larger networks. However, the output variance is small for such cases, and the slightly higher output variance with the larger networks was offset by the slightly higher noise variance found with the smaller networks.

4.2 Tests of the behaviour of large networks 109

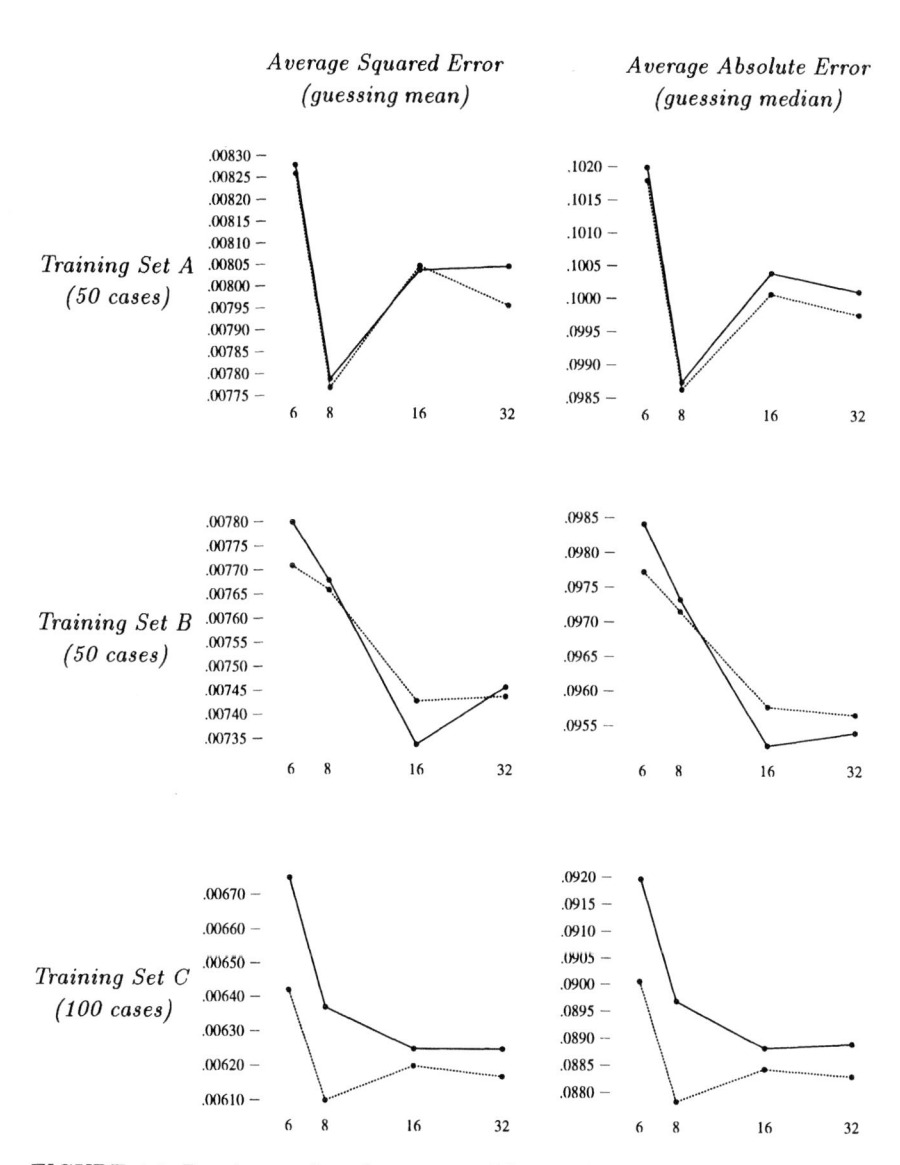

FIGURE 4.2. Results on the robot arm problem with networks of varying size. Three training sets were used, two with 50 cases, one with 100 cases. For each training set, Bayesian learning was done for networks with 6, 8, 16, and 32 hidden units, using both Gaussian priors for the hidden-to-output weights (solid lines) and Cauchy priors (dotted lines). Performance is shown in terms both of average squared error and of average absolute error, in each case when guessing optimally for that loss function. Performance was measured on a test set of 10 000 cases.

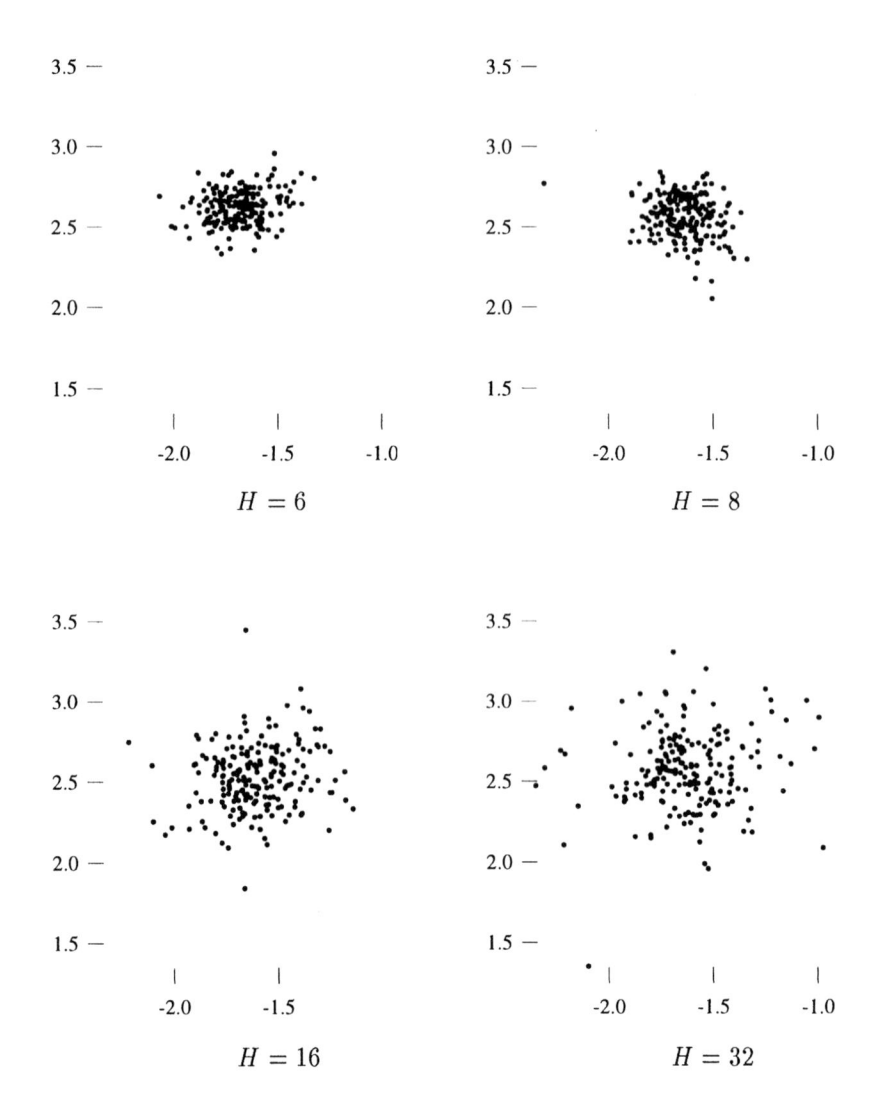

FIGURE 4.3. Predictive distributions obtained using networks of varying size. The plots show samples of 200 points from the predictive distributions based on training set A for the target values associated with inputs $(2.0, 0.5)$, a point just outside the region of the training data, as defined by network models with 6, 8, 16, and 32 hidden units, using Gaussian priors. The 200 points were obtained using the networks from the last 200 super-transitions of the Markov chain Monte Carlo runs.

4.2 Tests of the behaviour of large networks   111

a more restricted range of predictions than would a larger network, especially when extrapolating to regions where the training data is sparse. Since we will seldom have reason to believe that the true function is in the restricted class representable by a small network, the greater certainty of the predictions produced by the small networks will usually be unwarranted.

No clear difference in performance on this problem was seen between the networks using Gaussian priors for the hidden-to-output weights and those using Cauchy priors, though networks with different characteristics were found when the two different priors were used (in particular, the largest of the hidden-to-output weights tended to be larger in the networks learned with the Cauchy prior than in those learned with the Gaussian prior). This is disappointing, since as discussed in Section 4.2.1, one might sometimes expect to see such differences. The robot arm problem may perhaps be too simple to provide insight into this matter.

The good behaviour observed using Bayesian learning with large networks contrasts sharply with the behaviour of maximum likelihood training. This is illustrated in Figure 4.4, which shows results of maximum likelihood learning on training set A (consisting of 50 cases). For these tests, I fixed the priors for all the weight classes to be Gaussian with a standard deviation of 1000, and then found the maximum *a posteriori* probability (MAP) estimate for the network parameters. This is equivalent to maximum penalized likelihood estimation with a very small penalty; including this small penalty avoids the problem that the true maximum likelihood estimate could lie at infinity (though it is still possible that the true estimate lies sufficiently far out that it will not be found in a run of reasonable length). I let the standard deviation of the noise be determined by the data, as in the Bayesian runs. This has no significant effect on the location of the maximum, but does influence the progress of the maximization procedure.

Training for these tests was done using a method similar to the standard "backprop with momentum" technique (Rumelhart, Hinton, and Williams 1986b), which I implemented by suppressing the stochastic aspect of the dynamical techniques used for the Bayesian learning. (This is not necessarily the most efficient method, but it was convenient in this context.) The MAP estimate was found by repeatedly updating the network parameters and associated momentum variables via "leapfrog" steps (equations (3.9)–(3.11)), with each step being based on the full gradient computed using all training cases. The leapfrog stepsize was the same for all parameters, and was set manually, as the heuristic stepsize selection procedure relies on hyperparameter values that are not present in this context. After each leapfrog step, the momentum variables were multiplied by a factor, γ, less than one. For early iterations, Gaussian noise of variance $1 - \gamma^2$ was added to the momentum after the multiplication by γ, which has the effect of leading the system to an approximation to the Bayesian posterior distribu-

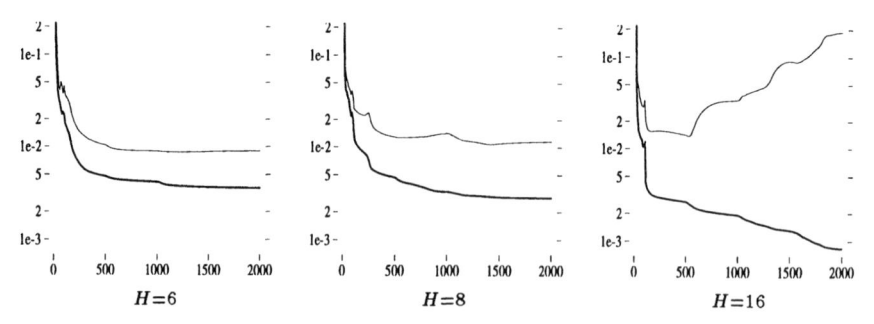

112   Chapter 4. Evaluation of Neural Network Models

H=6   H=8   H=16

FIGURE 4.4. Results of maximum likelihood learning with networks of varying size. Networks with 6, 8, and 16 hidden units were learned using training set A (containing 50 cases). The plots show the progress during learning of the average squared error on the training set (thick line) and on the test set (thin line). The horizontal axes gives the number of training iterations, in thousands, with points being plotted every 10 000 iterations.

tion. For later iterations, this noise was not added, causing the system to converge to a local maximum of the (slightly penalized) likelihood.[7]

The results in Figure 4.4 confirm the standard view that limiting the size of the network is necessary when learning is done using maximum likelihood. The network with 16 hidden units severely overfit the data — the squared error on the training set fell to an unrealistically low level, while the squared error on the test set became very bad. Indeed, this network had apparently not fully converged to the maximum likelihood solution; if it had, there is every reason to think that performance on the test set would have been even worse.

Performance with networks of 6 and 8 hidden units was better, but not as good as the performance using Bayesian learning. For the final parameter values, which appear as if they may be close to the true MAP estimates, the squared error on the test set was 0.00905 for $H = 6$ and 0.01155 for $H = 8$; for comparison, the worst performance of any of the Bayesian networks trained on this data set was 0.00828 (for $H = 6$, with a Gaussian prior).

The problem of overfitting can sometimes be alleviated by "early stopping" — halting training sometime before the maximum is reached, based on performance on a validation set separate from the training set (this is discussed, for instance, by Baldi and Chauvin (1991)). For a problem with as small a training set as considered here (50 cases), early stopping is

---

[7] In detail, the procedure was as follows: After initialization of the parameters to zero, there were 10 000 leapfrog steps with $\epsilon = 0.008$ and $\gamma = 0$, with noise added, then 100 000 steps with $\epsilon = 0.006$ and $\gamma = 0.9$, with noise added, then 400 000 steps with $\epsilon = 0.003$ and $\gamma = 0.99$, with no noise added, then 500 000 steps with $\epsilon = 0.002$ and $\gamma = 0.999$, with no noise added, and finally 1 000 000 steps with $\epsilon = 0.002$ and $\gamma = 0.9999$, with no noise added, for a total of 2 000 000 leapfrog steps.

probably not attractive, since setting aside some of this data for use as a validation set would likely degrade performance substantially. In any case, as can be seen from Figure 4.4, early stopping would have at best improved performance only slightly for the networks with 6 and 8 hidden units. For the network with 16 hidden units, early stopping could have been advantageous, but performance would still have been worse than with the smaller networks. (Note, however, that the effect of early stopping may depend a great deal on the particular optimization method used.)

Overfitting can also be addressed by adding a penalty term to the log likelihood, a procedure known as "weight decay" in the neural network context. In an earlier comparison (Neal 1993a), I found (in one case, at least) that weight decay can give results not much worse than are obtained using Bayesian learning, provided the right degree of weight decay is used. Determining the right degree of weight decay again requires a validation set, however, which will reduce the amount of data in the training set.[8]

To summarize, these tests support the conclusion that with Bayesian learning one can use a large network even when the training set is small, without overfitting. This result is of significant practical importance — when faced with a learning problem, we can simply use a network that is as large as we think may be necessary, subject to computational constraints, rather trying somehow to determine the "right" size of network. By not restricting the size of the network, we avoid the possibility that a small network might not produce as good predictions (seen in Figure 4.2 with respect to the networks with only 6 hidden units), as well as the possibility that a small network may produce overly-confident predictions (as illustrated in Figure 4.3). However, as indicated in Figure 4.1, training a large network can take a long time. In practice, though, the training time for a problem of this sort would usually not be quite this long — I have here been rather generous in the length of the runs in order to increase confidence that the results are based on the true equilibrium distributions.

4.3 Tests of Automatic Relevance Determination

The Automatic Relevance Determination (ARD) model developed by David MacKay and myself was described briefly in Section 1.2.3. Its aim is to automatically determine which of many inputs to a neural network are relevant to prediction of the targets. This is done by making the weights on the

---

[8] Alternatively, an n-way cross-validation scheme might be used, based on n divisions of the available data into training sets and validation sets. This is computationally expensive, however, and for neural networks might not work well in any case, due to the possibility that the networks found with different divisions may lie in dissimilar local minima.

connections out of each input unit have a distribution that is controlled by a hyperparameter associated with that input, allowing the relevance of each input to be determined automatically as the values of these hyperparameters adapt to the data.

I have tested the ARD model on the noisy LED display problem used by Breiman, et al (1984), and on a version of the robot arm problem with irrelevant inputs. The tests on the noisy LED display problem also allow an evaluation of how well a hierarchical model can adapt the architecture of a network to the data.

4.3.1 Procedures for evaluating ARD models

To evaluate the ARD model, a well-defined alternative is needed for comparison, for which the obvious choice is a model with a single hyperparameter controlling the weights on connections out of all input units. For some problems, this alternative will be ill-defined, however, since if the inputs have different dimensions, the results will depend on the arbitrary choice of measurement units. In such cases, it is necessary to adjust the scales of the inputs on the basis of prior knowledge, so as to make a one-unit change in one input have the same possible significance as a one-unit change in any other input.

Such prior knowledge may be helpful for the ARD model as well. When the ARD model is used in problems with many input variables, it may be necessary to use informative priors for the hyperparameters associated with the inputs. If vague priors are used for large numbers of hyperparameters, the prior probability of their taking on values appropriate for any particular problem will be very low, perhaps too low to be overcome by the force of the data. The posterior distribution of the hyperparameters may then be quite broad, rather than being localized to the appropriate region. Note, by the way, that this should not be a problem when single-valued estimates for the hyperparameters are used that maximize the probability of the data (the "evidence"), as is done by MacKay (1991, 1992b). A single-valued estimate is of course always localized, and the location of this estimate will usually not be affected by a widening in the permitted range of the hyperparameters. Consequently, one can avoid the effort of selecting informative priors when using this technique. Overall, this effect is not necessarily desirable, however, since there are presumably times when the posterior distribution of the hyperparameters *should not* be localized, but should instead be spread over a region whose extent is comparable to that of the correct informative prior.

Rather than use different informative priors for different input hyperparameters, we can instead use the same prior for all of them, after rescaling the inputs so that a one-unit change has similar significance for each, as described above. Even once this is done, however, there are still several apparently reasonable priors one might use. I consider two possibilities:

- A 1-level prior, in which the ARD hyperparameters are independently given rather vague priors.

- A 2-level prior, in which a high-level hyperparameter common to all inputs is given a very vague prior, while the ARD hyperparameters applying to each input are given less vague priors, with prior mean determined by the common high-level hyperparameter.

The second scheme is meant to avoid the possible problems with vague priors discussed above, but without fixing the overall degree of significance of the inputs, which may not be intuitively clear.

It is also desirable, with or without ARD, for the values of the inputs to be shifted so that the centre of the region of possible significance is zero. This is needed for it to be sensible to use a Gaussian of mean zero as the prior for the hidden unit biases.

Unfortunately, shifting and scaling the inputs according to prior knowledge as described above is not really possible for the tests done in this chapter. For the two synthetic data sets, we know exactly how the data was generated, and therefore could in theory figure out exactly how to rescale and shift the inputs to achieve optimal performance. This would say little about performance on real problems, however. I have therefore chosen to use the obvious forms of the inputs for these problems, which seem fairly reasonable.

For the other data sets, we have the problem that although the data is real, the context is now artificial. We no longer have access to whatever expert knowledge might have been used by the original investigators to rescale the inputs to equalize their potential relevance. On the other hand, we do know the results of past evaluations of other learning procedures applied to this data, which might allow this to be done in an unfair fashion.

I have handled this problem by "normalizing" the inputs in the real data sets — that is, by shifting and rescaling each input so as to make its mean be zero and its standard deviation be one across the training set. This is a common procedure, used by Quinlan (1993), for example.

From a Bayesian viewpoint, this normalization of inputs may appear to make little sense. In some cases, the values of the input variables are simply chosen by the investigator, in which case their distribution would seem to have nothing to do with the relationship being modeled. In other cases, the inputs have some distribution determined by natural variation, but the investigator's decisions heavily influence this distribution. In an agricultural field trial, for instance, the amount of fertilizer applied to each plot is just whatever the experimenter decides to apply. The distribution of

FIGURE 4.5. Digit patterns for the noisy LED display problem.

these decisions says something about the mental state of the experimenter, but it says nothing, one would think, about the effects of fertilizer on crop yield. The experimenter might also measure the amount of rainfall on each test plot. Though the rainfall is not determined by the experimenter, its distribution is heavily influenced by the experimenter's decision on how widely to disperse the test plots. Deciding to put some of the test plots in the desert, for example, would radically alter this distribution.

However, normalizing the inputs makes some degree of sense if we are willing to assume that the original investigators made sensible decisions. If so, they presumably arranged for the distribution of the input values to cover the range over which they expected significant effects, but did not wastefully gather data far beyond this range. If for some particular input they failed to gather data over what they considered an adequate range, they presumably omitted that input from the final set used. In the absence of any presently available expert opinion, these presumptions may be the best guide to the range over which each input might possibly have significance. Normalizing the inputs will then equalize these ranges, making use of a non-ARD procedure sensible, and allowing simple informative priors to be used in an ARD model.

4.3.2 Tests of ARD on the noisy LED display problem

The noisy LED display problem was used by Breiman, Friedman, Olshen, and Stone (1984) to evaluate their Classification and Regression Tree (CART) system. The task is to guess the digit indicated on a seven-segment LED display in which each segment has a 10% chance of being wrong, independently of whether any other segments are wrong. The correct patterns for the digits are shown in Figure 4.5. The ten digits occur equally often. The correct patterns, the frequencies of the digits, and the nature of the display's malfunction are assumed not to be known *a priori*.

The seven segments to be recognized are presented as seven input variables taking on values of "off" and "on", which I represent numerically as -0.5 and $+0.5$. (A symmetric representation seems appropriate, since the problem description contains no information regarding the meaning of "off" vs. "on", and since the CART system also treats the two possible values symmetrically.) In one version of the problem, only these seven inputs are present; in another version, seventeen addition irrelevant input variables

4.3 Tests of Automatic Relevance Determination

are included, each taking the values -0.5 and $+0.5$ with equal probability, independently of any other variable. The latter version will provide a test of the ARD model.

Breiman, *et al* randomly generated training sets of 200 examples, and tested performance of the resulting classifier on a test set of 5000 additional examples. They report that in these tests the classification tree produced by the CART system mis-classified about 30% of the test examples, regardless of whether the seventeen irrelevant attributes were included. The optimal classification rule based on knowledge of the true distribution of the data has a 26% mis-classification rate.

In applying a neural network to this ten-way classification problem, it is appropriate to use the "softmax" model (Bridle 1989), which corresponds to the generalized logistic regression model of statistics (see Section 1.2.1 and Section A.1.2). The network will take the values representing the seven segments along with any irrelevant attributes as inputs, and produce ten outputs, corresponding to the ten possbile digits. The conditional probability of a digit, given the inputs, is defined to be proportional to the exponential of the corresponding output.

This problem can be solved optimally by a network without any hidden units (only direct connections from input units to output units). There appears to be nothing in the problem description to indicate that a linear network would be adequate, however, so it might be regarded as unfair to take advantage of this fact. I therefore used networks containing a layer of eight hidden units, fully connected to the input units and to the output units. I did provide direct connections from inputs to outputs as well, so that a perfect solution was possible. I also trained networks without hidden units, to see whether such a restricted model actually did perform better.

The seven segments are equally relevant in this classification problem, in the sense that in a network implementing an optimal solution the weights from these seven inputs will all be of equal magnitude. The problem description does not indicate that these inputs are equally relevant, however, so again it might seem unfair to assume this when solving the version without irrelevant attributes. I therefore used an ARD model, with separate hyperparameters controlling the weights out of each input. When no irrelevant inputs are present, ARD might be detrimental, whereas when irrelevant attributes are present, ARD is expected to improve performance. For comparison, a model with a single hyperparameter controlling the weights from all inputs was tested on both version of the problem as well.

In all, four network architectures were tested — a network with no hidden units without ARD, a network with no hidden units with ARD, a network with a hidden layer without ARD, and a network with a hidden layer with ARD. The last of these is the architecture whose use is in accord with the prior knowledge presumed to be available for this artificial problem.

Hierarchical priors were set up as follows. In all architectures, the biases for the output units were considered one parameter group, as were the biases for the hidden units, and the weights from hidden units to output units, if hidden units were present. For the non-ARD models, the weights from the inputs to the outputs formed a single parameter group, as did the weights from the inputs to the hidden units, if present. For the ARD models, the weights from each input to the outputs formed separate groups, as did the weights from each input to the hidden units, if present. For all groups, the associated precision was given a prior as in equation (4.2), with $\omega = 1$ and $\alpha = 0.2$, except for the hidden-to-output weights, for which ω was set equal to the number of hidden units (eight), in accord with the scaling properties discussed in Chapter 2. (Subsequent experience on other data sets indicates that priors with $\alpha = 0.2$ may be less vague than is desirable, but I did not realize this when these tests were done. As discussed in Section 4.3.1, it may be best to use a 2-level prior for ARD hyperparameters, but this also was not tried in these tests.)

All four architectures were applied both to the version of the problem with only the seven relevant inputs, and to the version with 17 additional irrelevant inputs. For each of these eight combinations, three runs were done, using three different randomly generated training sets of 200 cases. The same three training sets were used for each network architecture; training sets with irrelevant attributes were obtained by adding irrelevant attributes to the three training sets with only relevant attributes. The same test set of 5000 cases was used to evaluate performance for all combinations. These commonalities permit more precise assessment of the effects of the variations.

The initial phase of each run consisted of 200 pairs of Gibbs sampling updates for hyperparameters and hybrid Monte Carlo updates for parameters. The trajectories used consisted of $L = 50$ leapfrog steps, done with a stepsize adjustment factor of $\eta = 0.4$; . The windowed variant of hybrid Monte Carlo was used, with accept and reject windows of $W = 5$ states. Partial gradients were not used for this problem, as they appeared to give only a small benefit. The computation time required for the initial phase varied from three to nine minutes, depending on the network architecture and on whether irrelevant inputs were included.

For the sampling phase, the hybrid Monte Carlo updates were done with $\eta = 0.4$, $L = 500$, and $W = 10$. Each sampling phase consisted of 150 supertransitions, with each consisting of ten pairs of Gibbs sampling and hybrid Monte Carlo updates. Computation time for the sampling phase varied with the network architecture and with whether irrelevant inputs were included, but did not vary substantially with whether ARD was used. The time required was 2.8 hours without hidden units and without irrelevant inputs, 6.0 hours with hidden units and without irrelevant inputs, 5.1 hours without hidden units and with irrelevant inputs, and 9.4 hours with hidden units and with irrelevant inputs.

For all runs, the sampling phase appeared to have converged within 80 super-transitions. The states saved at the end of each of the last 70 super-transitions were therefore used to make predictions. Convergence was faster and dependencies smaller for networks without hidden units; for these, an adequate sample could in fact have been obtained using substantially fewer super-transitions than were actually performed.

Figure 4.6 shows the performance of these models in terms of percent mis-classification, as measured on a test set of 5000 cases, for a standard error of $\pm 0.65\%$.[9] Comparisons of the results using neural network models with those using the CART classification tree procedure, measured on a different test set, are therefore significant (at the 5% level) only if the difference in performance is greater than about 2%. Since the same test set was used for all the neural network figures, comparisons of different neural network models may be significant with differences less than this (as discussed by Ripley (1994a)). Recall also that the same three training sets are used for all the neural network models.

The results when no irrelevant inputs are included are uniformly good. One would expect a penalty from using ARD when all inputs are of equal relevance, and from using a model with hidden units when the problem can be optimally solved without them, but clearly any such penalty is undetectably small in this context. Though the results for neural network models seem slightly better than those Breiman, et al (1984) found for CART, this difference is not statistically significant.

The results when irrelevant inputs are included are more interesting. CART's cross-validation-based tree pruning procedure manages to prevent these irrelevant inputs from being used, so that performance is unaffected by their presence. In contrast, the neural network models that did not use ARD performed poorly in the presence of irrelevant attributes. ARD was successful at largely suppressing the bad effects of including irrelevant inputs, though there appears to still be a small penalty, as one would expect. The differences seen between CART and the neural network models using ARD are again not statistically significant.

The effects of using a separate hyperparameter to control the standard deviations of weights out of each input in the ARD models are evident in

---

[9] If p is the true probability of mis-classification, the variance of the total number of errors on K test cases is $Kp(1-p)$, giving a standard error of $\sqrt{p(1-p)/K}$, which for $K = 5000$, $p \approx 0.3$ is about 0.0065.

Type of model	Relevant attributes only			Plus 17 irrelevant attributes		
No hidden layer						
Without ARD	28.2%	29.1%	28.7%	37.8%	38.2%	37.2%
With ARD	28.9%	29.0%	28.5%	29.8%	31.2%	31.0%
Eight hidden units						
Without ARD	28.3%	29.1%	29.1%	37.8%	37.6%	33.1%
With ARD	28.4%	29.6%	29.5%	30.4%	31.7%	33.0%
Classification tree	30%			30% 31% 30% 30% 31%		

FIGURE 4.6. Results on the noisy LED display problem. The figures for neural networks show percent mis-classification on a 5000 item test set for three runs with different training sets. Results are shown for four network models, applied both to data sets with only the seven relevant attributes, and to data sets with these plus 17 irrelevant attributes. Results for classification trees produced by the CART system are also shown, from the tests by Breiman, et al (1984, Section 3.5.1). One CART run is shown for the problem with no irrelevant attributes; five done with different training sets are shown for the problem with irrelevant attributes. The training and test sets used for the CART tests are not the same as those used for the neural network tests.

Figure 4.7, which shows the average squared magnitudes of weights out of each input, for networks with no hidden units, with and without ARD. (Recall that for the softmax model used here, each input is connected to the ten outputs associated with the ten classes.) When ARD is used, the magnitudes of weights on connections out of the relevant inputs are bigger, and the magnitudes of weights on connections out of the irrelevant inputs are smaller, than when ARD is not used.

One of the training sets with irrelevant inputs (the third) produced slightly puzzling results — with hidden units in the network, the mis-classification rate was about 33% regardless of whether or not ARD was used, better than performance of non-ARD networks on the other two training sets, but worse than performance of ARD networks on the other two training sets. The near-identical performance appears to be a coincidence, since the hyperparameter values for the two networks indicate that they are not at all similar. However, it may be that by chance the irrelevant inputs in this training set contained some pattern that induced unusual behaviour both with and without ARD. One would expect this to happen occasionally with small training sets.

For the models with hidden units, the standard deviation of the hidden-to-output weights (a hyperparameter) took on fairly small values between about 0.1 and 1.0, except for two of the networks not using ARD with irrelevant inputs present, and the one network using ARD discussed in the previous paragraph. This may have helped prevent the presence of hidden

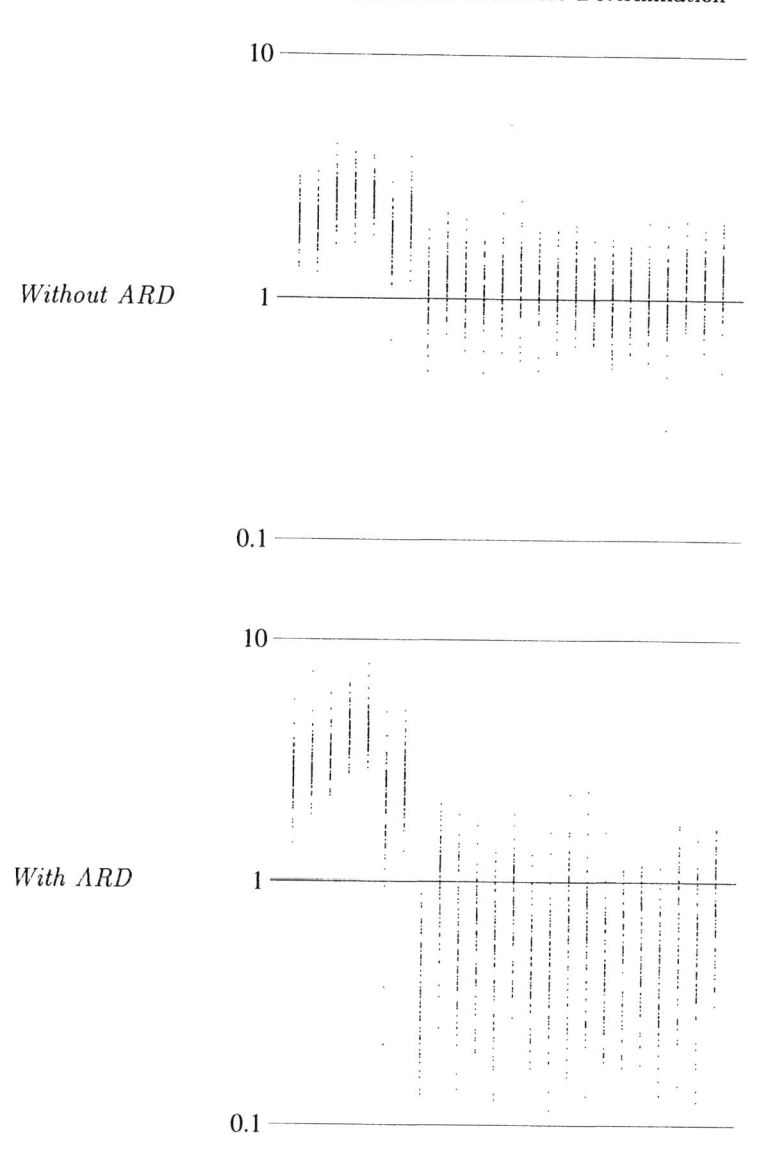

FIGURE 4.7. Relevant and irrelevant input weight magnitudes for the LED display problem. The plots show (on a log scale) the posterior distributions of the square roots of the average squared magnitudes of weights on connections out of each of the twenty-four inputs units, for the networks without hidden units, applied to the first training set. The first seven of these twenty-four inputs are relevant; the remaining seventeen are irrelevant. The plot on the left shows the magnitudes when ARD is not used (only one hyperparameter); that on the right shows the magnitudes when ARD is used (one hyperparameter for each input unit). The points in the plots were computed from the states saved after the last 80 super-transitions of the sampling phase.

units from having a damaging effect. One might actually have expected the standard deviation for these weights to take on even smaller values, effectively eliminating the hidden layer. That this did not happen may be due to the prior for the associated hyperparameter not being vague enough. Alternatively, the good test performance seen (with the ARD models) may indicate that these weights were sufficiently small as is.

4.3.3 Tests of ARD on the robot arm problem

I have also tested the Automatic Relevance Determination model on a variation of the robot arm problem. In this variation, six input variables, $x'_1, \ldots, x'_6$, were present, related to the inputs of the original problem, x_1 and x_2, as follows:

$$\begin{aligned} x'_1 &= x_1, & x'_3 &= x_1 + 0.02\, n_3, & x'_5 &= n_5 \\ x'_2 &= x_2, & x'_4 &= x_2 + 0.02\, n_4, & x'_6 &= n_6 \end{aligned} \quad (4.3)$$

where n_3, n_4, n_5, and n_6 are independent Gaussian noise variables of mean zero and standard deviation one. As in the original version, the targets were functions of x_1 and x_2 (equivalently, of x'_1 and x'_2), plus noise of standard deviation 0.05 (see equation (3.34)). Clearly, x'_5 and x'_6 are irrelevant to predicting the targets. In isolation, x'_3 and x'_4 would convey some information about the targets, but in the presence of x'_1 and x'_2, which contain the same information in noise-free form, they are useless, and should also be ignored.

We would like to see whether the ARD model can successfully focus on only x'_1 and x'_2, and, if so, whether this does indeed improve predictive performance. To test this, I generated new versions of the training set of 200 cases used before by MacKay (1991, 1992b) and for the demonstration in Section 3.3, and of the test set of 10 000 cases used in Section 4.2.2. The input variables in these data sets were derived from the corresponding inputs in the original data sets in accord with equation (4.3); the targets were the same as before. A model that completely ignored the irrelevant inputs would therefore be able to achieve the same performance when trained on this data as would a model trained on the original data without the irrelevant inputs.

For these tests, I used a network with a single hidden layer of $H = 16$ tanh units. For all models, the hidden-to-output weights were given Gaussian priors, whose precision was a common hyperparameter, to which I gave a vague Gamma prior (equation 4.2)) with $\omega = 100H$ and $\alpha = 0.1$. The hidden unit biases were also given Gaussian priors, with their precision being a hyperparameter that was given a Gamma prior with $\omega = 100$ and $\alpha = 0.1$. The output unit biases were given a Gaussian prior with a fixed standard deviation of one.

4.3 Tests of Automatic Relevance Determination

Gaussian priors were also used for the input-to-hidden weights, but the precisions for these Gaussian priors were specified in three different ways, to produce a non-ARD model, a 1-level ARD model, and a 2-level ARD model. In the non-ARD model, all the input-to-hidden weights had prior precisions given by a single hyperparameter, to which I gave a vague Gamma prior with $\omega = 100$ and $\alpha = 0.1$. For the 1-level ARD model, each input had an associated hyperparameter that controlled the prior precision of weights out of that input, with these hyperparameters being given independent Gamma priors with $\omega = 100$ and $\alpha = 0.1$. For the 2-level ARD model, each input again had its own hyperparameter, but these low-level hyperparameters were given somewhat less vague Gamma priors, with $\alpha = 0.5$, and with the mean ω being a common high-level hyperparameter. This high-level hyperparameter was given a very vague prior with $\omega = 100$ and $\alpha = 0.001$.[10]

For all three models, learning began with a short initial phase, and continued with a long sampling phase, consisting of 200 super-transitions. Each super-transition consisted of ten pairs of Gibbs sampling updates for hyperparameters and hybrid Monte Carlo updates for parameters. The hybrid Monte Carlo trajectories were $L = 4000$ leapfrog steps long, and were computed using partial gradients, based on a four-way division of the training set, with a stepsize adjustment factor of $\eta = 0.6$. The windowed acceptance procedure was used, with windows of ten states. Each of these runs required 42 hours of computation time.

Figure 4.8 shows how the magnitudes of weights on connections out of the different inputs varied in the course of the simulations for the three models. With both ARD models, the weights on connections out of the four irrelevant inputs quickly became a factor of ten or more smaller in magnitude than the weights on connections out of the two relevant inputs. The differences in average weight magnitudes for the model without ARD were considerably smaller (less than a factor of two).

It is interesting to compare the results seen with the 1-level ARD model to those seen with the 2-level ARD model. Although the Gamma prior used for the hyperparameters in the 1-level model was rather vague, it seems that it was not so vague as to have no influence on the results — the prior seems to have prevented the weight magnitudes for the irrelevant inputs from becoming much smaller than 0.01. The magnitudes for the weights from relevant input in the 1-level model are somewhat larger than in the 2-level model, perhaps due to residual pressure to increase the disparity with the weights from the irrelevant inputs. Since the prior for the low-

---

[10] This prior is perhaps vaguer than is necessary, but using a very low value for α has the advantage that it increases the acceptance rate of the rejection sampling scheme used to implement Gibbs Sampling for the high-level hyperparameter (see Appendix A, Section A.5).

124   Chapter 4. Evaluation of Neural Network Models

No ARD

1-level ARD

2-level ARD

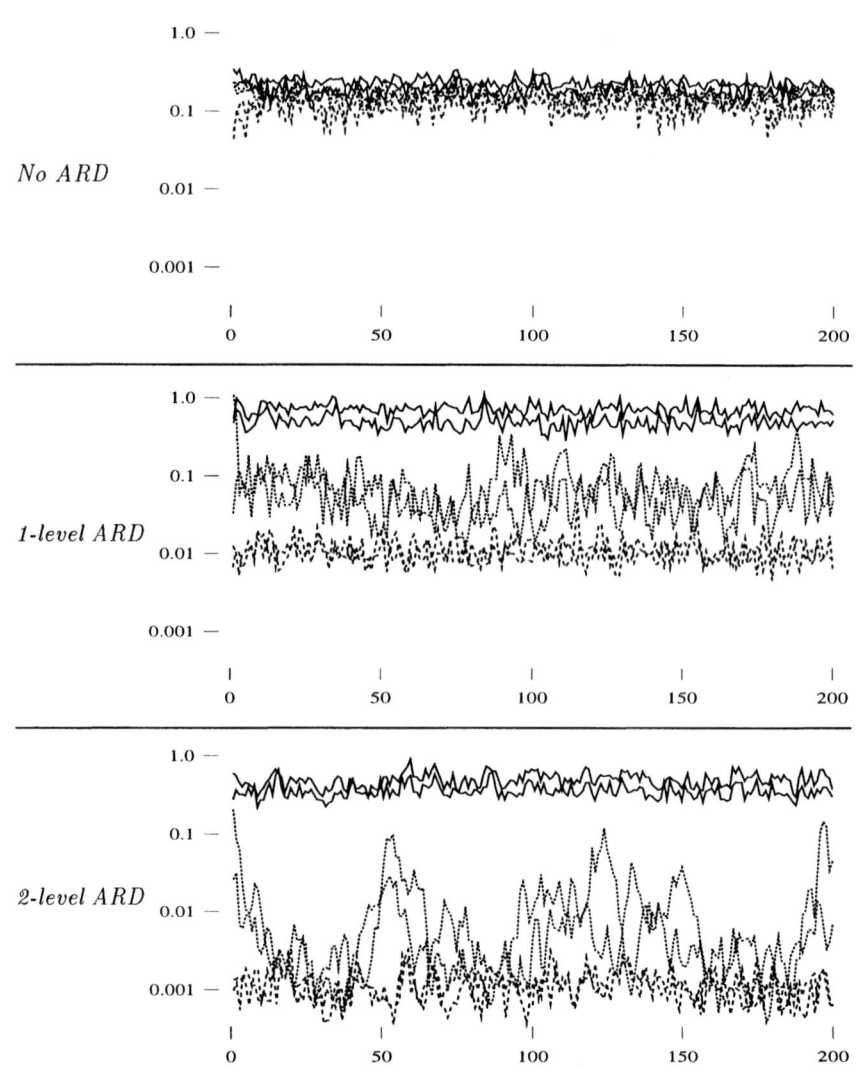

FIGURE 4.8. Input weight magnitudes for the robot arm problem with and without ARD. These plots show the square roots of the average squared magnitudes of weights on connections from the six inputs, with the magnitudes for the two inputs carrying noise-free information given by solid lines, the magnitudes for the two inputs carrying noisy information by the dotted lines, and the magnitudes for the completely irrelevant noise inputs by dashed lines (all on a log scale). The horizontal axis gives the number of super-transitions.

4.3 Tests of Automatic Relevance Determination 125

level hyperparameters in the 2-level model is less vague than that in the 1-level model, one might wonder why the weight magnitudes influenced by these hyperparameters were able to become more widely spread in the 2-level model. This is due to the asymmetry of the Gamma prior used, under which the upper tail for $\log \sigma$ is heavier than the lower tail. In the 2-level model, the mean for the low-level hyperparameters is a high-level hyperparameter with a very vague prior that allows it to adopt a value that positions the low-level prior where the heavy upward tail covers the appropriate range.

The difference seen between the 1-level and 2-level models is thus in part due to the particular form I have used for the priors. I expect that a 2-level model will have more general advantages, however. It may be dangerous to give very vague priors to many hyperparameters, since the prior probability of their taking on values matching the data will then be very small. In the 2-level model, only one high-level hyperparameter is given a very vague prior; the others have less vague priors that should nevertheless be adequate to permit the desired variation in weight magnitudes, once these priors are properly positioned by the adaptation of the high-level hyperparameter.

Since the ARD models succeeded in suppressing the weights on connections to irrelevant inputs, whereas the non-ARD model did not, one would expect that the predictive performance of the ARD models would be better than that of the non-ARD model. This was indeed the case. On the test set of 10 000 cases, the average squared error when guessing the average outputs of the networks from the last 150 super-transitions was 0.00597, 0.00563, and 0.00549 for the non-ARD model, 1-level ARD model, and 2-level ARD model, respectively.[11] The error using the 2-level ARD model was almost identical to the error of 0.00547 measured on this test set using the networks from the last 100 super-transitions of the first run with $L = 8000$ described in Section 3.3, which was trained on the same set of cases, but without the irrelevant inputs. It turns out that very similar predictive performance can be obtained from shorter runs — using only the last 50 of the first 75 super-transitions in these runs, the average squared error was 0.00589, 0.00564, and 0.00552 for the non-ARD, 1-level ARD, and 2-level ARD models. Runs of this length would take 16 hours of computation time.

---

[11] The standard errors for these figures are approximately ±0.00006, so the advantage seen for ARD is statistically significant with respect to the variation due to the random choice of test set. Since only a single training set was used, the random variation due to this factor cannot be quantified.

4.4 Tests of Bayesian models on real data sets

The tests on synthetic data sets described in the previous sections have helped clarify the properties of the network models and priors tested. This knowledge should be of use when appling Bayesian learning to real problems. In this section, I test the Bayesian network models and the Markov chain Monte Carlo implementation on two real data sets, one for a regression problem and one for a classification problem.

4.4.1 Methodology for comparing learning procedures

In comparing learning procedures, we may be interested in how they differ in many respects, including the accuracy of the predictions made, the amount of computation required to produce these predictions, the ease with which the problem can be formulated in an appropriate form, and the extent to which the construction of a predictive model increases our understanding of the nature of the problem. Only in the context of a real application will we be able to judge the relative importance of these aspects, and only in such a context will some of them be testable. Traditionally, neural networks and other machine learning procedures have been compared primarily on the basis of their predictive performance, with some attention also paid to their computational requirements, and these aspects have been tested using data which may be real, but for which the original context is no longer available. Despite its limitations, this is the approach I will take here.

Learning procedures cannot be compared in a complete absence of context, however. We must postulate some loss function in terms of which the quality of the predictions can be judged. Furthermore, for the results of a comparison to be meaningful, we must somehow distinguish between procedures which just happen to do well on a particular problem, and those which not only do well, but also might have been chosen prior to our seeing the test results for the various procedures. Which procedures might reasonably have been chosen will depend on what background knowledge is assumed to be available. For these sorts of tests, there is an implicit assumption that the background knowledge is very vague (but this is not quite the same as a complete absence of background knowledge).

For example, suppose we are comparing neural networks with other methods on two problems. On problem A, a neural network with one hidden layer does better than any other method. On problem B, a neural network with two hidden layers performs best. It would not be valid to claim that these results demonstrate the superiority of neural networks unless there was some way that the user could have decided on the basis of background knowledge and the training data alone that a network with a single hidden layer was appropriate for problem A, but one with two hidden layers was appropriate for problem B.

To lessen the potential for criticism on this basis, I have used hierarchical models that are capable of reducing to simpler models depending on the settings of hyperparameters. In networks with two hidden layers, for example, I include direct connections from the inputs to the second hidden layer, and use a hyperparameter that controls the magnitude of the weights from the first to the second hidden layer. If this hyperparameter takes on a very small value, the network will effective have only a single hidden layer. This idea was used earlier in the tests on the noisy LED display problem (Section 4.3.2); the ARD model can also be seen as an instance of this approach. An alternative is to somehow choose between discrete model alternatives on the basis of the training data. Bayesian methods for this are emphasized by MacKay (1992a), but the required computations are difficult in a Monte Carlo implementation (Neal 1993b, Sections 2.3 and 6.2). It is also possible to choose between models by other means, such as cross validation. Any of these methods may allow the effective model used to be determined to a large degree by the data. If the chosen model performs well, one can then argue that such good performance could indeed have been achieved in a real application of a similar nature.

4.4.2 Tests on the Boston housing data

The Boston housing data originates with Harrison and Rubinfeld (1978), who were interested in the effect of air pollution on housing prices.[12] The data set was used to test a method for combining instance-based and model-based learning procedures by Quinlan (1993). Although the original objective of Harrison and Rubinfeld was to obtain insight into factors affecting price, rather than to make accurate predictions, my goal here (and that of Quinlan) is to predict housing prices based on the attributes given, with performance measured by either squared error loss or absolute error loss.

The data concerns the median price in 1970 of owner-occupied houses in 506 census tracts within the Boston metropolitan area. Thirteen attributes pertaining to each census tract are available for use in predicting the median price, as shown in Figure 4.9. The data is messy in several respects. Some of the attributes are not actually measured on a per-tract basis, but only for larger regions. The median prices for the highest-priced tracts appear to be censored.[13]

---

[12] The original data is in StatLib, available via the World Wide Web, at URL http://lib.stat.cmu.edu/, under "datasets".

[13] Censoring is suggested by the fact that the highest median price of exactly $50,000 is reported for sixteen of the tracts, while fifteen tracts are reported to have median prices

CRIM	per capita crime rate by town
ZN	proportion of residential land zoned for lots over 25,000 sq.ft.
INDUS	proportion of non-retail business acres per town
CHAS	Charles River dummy variable (1 if tract bounds river, 0 if not)
NOX	nitric oxides concentration (parts per 10 million)
RM	average number of rooms per dwelling
AGE	proportion of owner-occupied units built prior to 1940
DIS	weighted distances to five Boston employment centres
RAD	index of accessibility to radial highways
TAX	full-value property-tax rate per $10,000
PTRATIO	pupil-teacher ratio by town
B	$1000\,(\text{Blk} - 0.63)^2$ where Blk is the proportion of blacks by town
LSTAT	percent lower status of the population

FIGURE 4.9. Descriptions of inputs for the Boston housing problem.

Considering these potential problems, it seems unreasonable to expect that the distribution of the target variable (median price), given the input variables, will be nicely Gaussian. Instead, one would expect the error (noise) distribution to be heavy-tailed, with a few errors being much greater than the typical error. To model this, I have used a t-distribution as the error distribution, as described in Appendix A, Section A.1.2. This is a common approach, used by Liu (1994), for example. I rather arbitrarily fixed the degrees of freedom for the t-distribution to the value 4. Ideally, one would let the degrees of freedom be a hyperparameter, but this is not supported by the present implementation.

Harrison and Rubinfeld (1978) consider various nonlinear transformations (e.g. logarithmic) of the target and input variables as the basis for their linear model. However, Quinlan (1993) uses only a linear transformation of the variables. Since I would like to compare with the results Quinlan gives, I did the same. A neural network should be able to implement whatever nonlinear transformation may be required, given enough data to go on, so modeling the untransformed data is a reasonable demonstration task. However, it seems likely that leaving the target (the median price) in its original form will result in the noise variance varying with the target value (heteroscedasticity). The procedures used by Quinlan apparently did nothing to adjust for this; neither do the neural network models I used, though it should be possible to extend them to do so. I expect that ignoring heteroscedasticity will degrade performance somewhat, but will not cause serious problems.

I did linearly transform the input variables and targets to normalize them to have mean zero and standard deviation one, as did Quinlan (1993).

---

above $40,000 and below $50,000, with prices rounded only to the nearest hundred. Harrison and Rubinfeld (1978) do not mention any censoring.

As discussed in Section 4.3.1, I view this procedure as a substitute for using expert knowledge to shift and rescale the input variables in order to equalize their potential relevance. For this data set, one way in which the prior knowledge of the original investigators may appear in the distribution of the input variables is through their selection of the study area — presumably Harrison and Rubinfeld believed that the range of variation in input variables seen over the Boston area was similar to the range over which these variables might be relevant, as otherwise they might have chosen to study housing prices in all of Massachusetts, or in just the suburb of Newton.

Quinlan (1993) assesses the performance of various learning procedures on this problem using ten-way cross validation. In this assessment method, each learning procedure is applied ten times, each time with nine-tenths of the data used for training and one-tenth used for testing, and the test errors for these ten runs are then averaged. Quinlan has kindly provided me with the ten-way division of the data that he used for his assessments.[14]

Since these cross validation assessments are computationally expensive, before undertaking any of them, I first evaluated a number of Bayesian neural network models using half the data (randomly selected) as a training set and the other half as a test set. These training and test sets both consisted of 253 cases.[15]

For the first of these preliminary tests, I trained a network with no hidden units, corresponding to a linear regression model. Since there is only one connection for each input in this model, ARD was not used — the input-to-output weights were simply given Gaussian distributions, with the precision for these Gaussian distributions being a common hyperparameter, which was given a Gamma prior with $\omega = 100$ and $\alpha = 0.1$ (see equation 4.2). The output bias was given a fixed Gaussian prior with standard deviation one. The noise distribution was a t-distribution with four degrees of freedom (see equation (A.6) in Appendix A), with the associated precision, σ^{-2}, having a Gamma prior with $\omega = 100$ and $\alpha = 0.1$.

This simple network was trained for 100 super-transitions, each consisting of ten pairs of Gibbs sampling and hybrid Monte Carlo updates. Trajectories were 100 leapfrog steps long, with a stepsize adjustment factor of 1.0. Total training time was seven minutes.

The states saved after the last 50 of these super-transitions were used for prediction. The resulting performance is reported in Figure 4.10, along

---

[14] This division of the data is stratified by target value, as described by Breiman, et al (1984, Section 8.7.2).

[15] In these tests, I used a slightly incorrect normalization procedure, which has the effect of adding a small amount of random noise to the inputs. This was fixed for the later cross-validation assessments, and turns out to have had little effect in any case.

130   Chapter 4. Evaluation of Neural Network Models

Model or procedure used	Average squared error	Average absolute error	Average neg log prob
Guessing mean of training set	83.4	6.70	–
Guessing median of training set	82.4	6.40	–
Network with no hidden units	28.9	3.36	2.888
Network with 8 hidden units			
With Gaussian prior	13.7	2.32	2.428
With Cauchy prior	13.1	2.26	2.391
Network with 14 hidden units			
With Cauchy prior	13.5	2.29	2.407
Network with two hidden layers	12.4	2.15	2.303

FIGURE 4.10. Results of preliminary tests on the Boston housing data. The predictions for each network model were based on the posterior distribution given the training set of 253 cases, as sampled by the Markov chain simulations. The figures are averages over the 253-case test set of the squared error when guessing the predictive mean, the absolute error when guessing the predictive median, and the negative log probability density of the true target value.

with that of the other networks trained in the preliminary tests, to be discussed shortly. Three performance criteria are used here — average squared error on the test set, when guessing the mean of the predictive distribution; average absolute error, when guessing the median of the predictive distribution; and average negative log probability density of the actual target value under the predictive distribution. Squared error can be very sensitive to a small number of large errors; absolute error is less so; negative log probability density is perhaps the best indicator of overall performance when there are occasional large errors.

Next, I trained networks with a layer of eight hidden units, using both Gaussian and Cauchy priors for the hidden-to-output weights. For these networks, I used a 2-level ARD prior for the input-to-hidden weights, with $\omega = 100$, $\alpha = 0.001$ for the high-level Gamma prior (for the common hyperparameter), and $\alpha = 0.5$ for the low-level Gamma prior (for the hyperparameters associated with particular inputs). The prior for hidden biases was Gaussian, with the precision having a Gamma prior with $\omega = 100$ and $\alpha = 0.1$. In the Gaussian network, the prior for hidden-to-output weights was Gaussian with a precision that I gave a Gamma prior with $\omega = 100H$ and $\alpha = 0.1$. Here H is the number of hidden units, here eight; this give proper scaling with network size, as discussed in Chapter 2. In the Cauchy network, a Cauchy prior for the hidden-to-output weights was implemented using a 2-level Gaussian prior, with $\omega = 100H^2$, $\alpha = 0.1$ for the high-level Gamma prior (for the common hyperparameter), and $\alpha = 1$ for the low-level Gamma prior (for the hyperparameters associated with particular hidden units).

I included direct connections from the inputs to the outputs in these networks. The weights on these direct connections, the bias for the output unit, and the level of the noise were all given the same priors as for the network with no hidden units.

Following a relatively short initial phase, these networks were trained for 250 super-transitions, each super-transition consisting of ten pairs of Gibbs sampling and hybrid Monte Carlo updates. The states at the ends of each of the last 150 super-transitions were used to make predictions. Trajectories were 1500 leapfrog steps long, with a stepsize adjustment factor of 0.6. They were computed using partial gradients, with a five-way division of the training data. The windowed acceptance procedure was used, with a window size of ten. Total training time was 21 hours for each network.

As can be seen in Figure 4.10, the networks with eight hidden units performed much better than the network with no hidden units. The results observed using the Cauchy prior were slightly better than those observed using the Gaussian prior, but the difference should probably not be regarded as significant.

Finally, I trained two more complex networks: one with a single hidden layer of fourteen hidden units, another with two hidden layers, each of six hidden units. In both networks, the hidden and output layers had direct connections to the inputs. These networks both had 224 parameters (weights and biases).

The priors used for the network with a single layer of fourteen hidden units were the same as for the network with eight hidden units, using the Cauchy prior (except for difference due to the scaling with H). The network was also trained in the same way as were those with eight hidden units, except that a longer initial phase was used, and the sampling phase was continued for 300 super-transitions, with the states saved from the last 200 being used for predictions. Total training time was 46 hours.

For the network with two hidden layers, I used a Gaussian prior for weights from the first hidden layer to the second hidden layer, and a Cauchy prior for weights from the second hidden layer to the outputs. This choice was inspired by Figure 2.10, which shows interesting two-dimensional functions produced from a similar model that combines Gaussian and non-Gaussian priors. (However, one may doubt whether six is really close enough to infinity for this picture to be relevant. Such priors may also behave differently with thirteen inputs than with two.)

In detail, the network model with two hidden layers used the following priors. For the weights on connections from the inputs to the first hidden layer, a 2-level ARD prior was used with $\omega = 100, \alpha = 0.1$ for the high-level

Gamma prior, and $\alpha = 3$ for the low-level Gamma prior. An ARD prior of the same form was used for the weights on connections from the inputs to the second hidden layer. The choice of $\alpha = 3$ for the low-level Gamma prior produces a distribution that is not too broad; I chose this somewhat narrow prior primarily to avoid any possible problem with the simulation becoming lost for an extended period in some strange region of the hyperparameter space. The weights on connections from the first hidden layer to the second hidden layer were given Gaussian priors, with precisions given by a Gamma prior with $\omega = 100 H_1$ and $\alpha = 0.1$, where $H_1 = 6$ is the number of units in the first hidden layer. For the weights on connections from the second hidden layer to the outputs, I implemented a Cauchy prior using a 2-level Gaussian prior with $\omega = 100 H_2^2$, $\alpha = 0.1$ for the high-level Gamma prior, and $\alpha = 1$ for the low-level Gamma prior, where $H_2 = 6$ is the number of units in the second hidden layer. The priors on the biases for the two hidden layers were both Gausian, with precisions given by Gamma priors with $\omega = 100$ and $\alpha = 0.1$. The priors for the input-to-output weights, the output biases, and the noise level were the same as for the other networks.

Training for the network with two hidden layers began with a short initial phase, which was followed by 100 super-transitions using the same learning parameters as were used for the networks with one hidden layer. In the last twenty of these super-transitions, the rejection rate climbed to over 50%. I therefore reduced the stepsize adjustment factor from 0.6 to 0.45, and increased the trajectory length from 1500 to 2000 to compensate. With these parameters, I let the run continue for another 200 super-transitions. The states from these 200 super-transitions were the ones used for predictions. Total training time was 54 hours.

As can be seen in Figure 4.10, the performance of the network with a single layer of fourteen hidden units differed little from that of the networks with only eight hidden units. However, performance of the network with two hidden layers did appear to be better than that of the networks with only one hidden layer.

Following these preliminary runs, I decided to do a cross-validation assessment of the network with two hidden layers (each with six hidden units), in order to compare with the results reported by Quinlan (1993). Technically speaking, this is cheating — this network architecture was chosen with knowledge of results involving all the data, whereas training for each component of the cross-validation assessment is supposed to be based solely on the nine-tenths of the data allocated to training for that component. There are two reasons why this does not invalidate the results. First, one could apply the same methodology of selecting an architecture (using preliminary runs trained with a subset of the data) within each component of the cross-validation assessment. Since the training and test sets for these runs would be only slightly smaller than for the preliminary runs done here, the results would likely be similar. (This was not done because it would have required considerably more computation time.) Second, the network architecture selected is that which is the most complex, the one that would be selected *a priori* under the philosophy of modeling that I am advocating. The preliminary runs simply confirm that, as expected, using a simpler architecture is not advantageous.

The objective of the assessments that Quinlan (1993) reports was to evaluate whether his scheme for combining "instance-based" and "model-based" learning was beneficial. Instance-based methods (such as k-nearest neighbor) make predictions based on similarities with "prototype" patterns. Model-based methods (such as neural networks) may use more general representations of regularities. Quinlan proposes a combined scheme in which a prediction for a particular test case is obtained by applying the instance-based method after adjusting the values associated with each prototype by the amount that the model-based method predicts the prototype's value will differ from that of the test case.

For my purposes, Quinlan's results simply indicate the performance achievable by reasonably sophisticated applications of existing techniques, thereby providing a standard against which I can compare the performance obtained with a Bayesian neural network model. The neural network component of Quinlan's assessment was done by Geoffrey Hinton. The network he used had a single hidden layer, and was trained to minimize squared error on the training set plus a weight decay penalty. The number of hidden units and the amount of weight decay were chosen by cross validation. In principle, this choice would be made ten times, once for each component of the main cross-validation assessment, but to save time a single choice was made. The network chosen in this way had fourteen hidden units (Geoffrey Hinton, personal communication).

I estimated that a ten-way cross-validation assessment of the Bayesian network model with two hidden layers that used the same training procedure as in the preliminary runs would required a total of 41 days of computation time. Wishing to reduce this, I performed a number of tests using states from the preliminary run. In particular, I looked at the correlations of various quantities along trajectories, in order to select a good trajectory length, and at the change in free energy from start to end of a trajectory when using various stepsize adjustment factors, window sizes, and partial gradient divisions, in order to select trajectory computation parameters that would give a good acceptance rate at minimal cost.

Based on these tests, I chose the following three-phase training procedure for use in the cross-validation assessment. Starting with weights and biases set to zero, I first trained the network for 1500 pairs of Gibbs sampling and hybrid Monte Carlo updates, using trajectories 100 leapfrog steps long (with a window of 10 states), with a stepsize adjustment factor of 0.5.

Model or procedure used	Ave sqr error	Ave abs error
Guessing overall mean	84.4	6.65
Guessing overall median	86.2	6.53
Bayesian neural network		
With no hidden units	25.3	3.20
With two hidden layers*	6.5	1.78
Instances alone	19.2	2.90
Max. likelihood linear regression	24.8	3.29
+ instances	14.2	2.45
Model tree	15.7	2.45
+ instances	13.9	2.32
Neural network using cross validation	11.5	2.29
+ instances	10.9	2.23

*Performance on each of the ten divisions: Squared error: 6.4, 7.0, 5.3, 10.0, 4.4, 6.0, 13.2, 3.6, 4.8, 3.9
Absolute error: 1.78, 1.87, 1.81, 2.13, 1.47, 1.78, 2.43, 1.38, 1.60, 1.49.

FIGURE 4.11. Cross-validation assessments on the Boston housing data. The figures are averages of performance (in terms of squared and absolute error) over all ten divisions of the data into training and test sets (except for the figures using overall means and medians, for which this would not be meaningful, due to stratification). The results in the bottom section are as reported by Quinlan (1993).

Next, I did 750 pairs of updates using trajectories 200 leapfrog steps long (with a window of 20 states), with a stepsize adjustment factor of 0.6. Finally, I ran a sampling phase consisting of 120 super-transitions, each consisting of ten pairs of Gibbs sampling and hybrid Monte Carlo updates, using trajectories 1000 leapfrog steps long (with a window of 30 states), with a stepsize adjustment factor of 0.6. The states saved after the last 100 of these super-transitions were used to make predictions. Trajectories in all phases were computed using the partial gradient method, with a ten-way division of the training data. Total training time was 27 hours for each network, 270 hours for the entire assessment.

The results of this assessment and those of Quinlan are shown in Figure 4.11.[16] As a check, I also did a cross-validation assessment of the network with no hidden units; as expected, its performance is similar to that which Quinlan reports for linear regression.

---

[16] Note that Quinlan reports squared error in terms of "relative error" with respect to the squared error guessing the overall mean of the data. To convert his results to the form displayed, multiply by 84.4.

4.4 Tests of Bayesian models on real data sets 135

The Bayesian neural network model with two hidden layers of six units performed substantially better than any of the other methods. To do a formal test for the significance of the difference in average performance seen, one would need the individual results for the other methods on each of the ten divisions of the data.[17] The individual results for the Bayesian network with two hidden layers are given at the foot of the figure. Unless the other methods exhibit greater variability in performance over the ten divisions than is the case for the Bayesian network model, it seems fairly implausible that the difference seen could be due to chance.

What is responsible for the good performance seen with this Bayesian network model, particularly as compared with the neural network trained by standard methods? Several aspects of the model might be important: the use of a network with two hidden layers, the use of an ARD prior, the use of a t-distribution for the noise, and the use of direct connections from inputs to all layers. The fact that the Bayesian training procedure averages the results of many networks might be crucial. The Markov chain Monte Carlo implementation might also be better at escaping local minima than the minimization procedure used for the standard network training.

I have not attempted to isolate all these possible influences. I did train a network of the same architecture (two hidden layers of six units each) to minimize the standard criterion of squared error, plus a small weight decay penalty, and found that serious overfitting occurred. Even stopping training at the optimal point as judged by the test set gives performance of only 9.7 in terms of squared error and 2.19 in terms of absolute error. This is slightly better than the other non-Bayesian methods, but not close to the performance of the Bayesian network. Of course, it is cheating to pick the stopping point using the test error, so the actual performance achievable with this procedure would be somewhat worse. On the other hand, choosing a better weight decay penalty by cross-validation might have improved performance.

I will also note a few relevant features of the posterior distributions found in the training runs that may shed some light on the reasons for the good performance seen. The weights on the direct connections from inputs to outputs were typically small, less than 0.1, but not completely negligible. Weights on direct connections from inputs to the second hidden layer were also mostly small, usually less than 0.5, except for the weights on connections from the DIS input, which often exceeded one. Weights on connections from the inputs to the first hidden layer were much larger,

---

[17] Even with this information, such a test might be problematical, since the distribution of performance for a method cannot be assumed to be Gaussian, or even unimodal, and since the ten performance values obtained in a cross-validation assessment such as this are not independent.

typically greater then one, and sometimes greater than ten. In many of the runs, such weights were substantially larger on connections from a few of the inputs than on the other connections. These features indicate that the first hidden layer is playing an important role in the network, and that the use of hyperparameters, and of ARD priors in particular, may have been beneficial.

In my view, the results of this test can be taken as evidence of the benefit of the Bayesian approach regardless of what particular modeling choices may have been responsible for the performance improvement. Ultimately, we are interested in the overall merits of different problem-solving methodologies, which, among other things, determine how such modeling choices are made. The Bayesian approach is based on probabilistic modeling of relationships, in which it is natural, for instance, to use a t-distribution for the noise whenever that seems appropriate, regardless of what loss function will be associated with the final predictions. In some other approaches, the fact that performance will ultimately be judged by squared error would lead to the use of squared error as a fitting criterion during training as well. In the Bayesian approach, we also need not fear overfitting, and hence are free to use a network with many parameters and a complex structure whenever it seems that the consequent flexibility may be useful. It is possible that techniques such as weight decay might be able to control overfitting by such a complex network when trained by non-Bayesian methods, but assurances of this are lacking. Consequently, users of a non-Bayesian methodology may choose an overly-simple model, out of fear of overfitting, even when a more complex model would in fact have worked well.

On the other hand, these tests show that there is a considerable need for improvement with respect to the computation time required by the Markov chain Monte Carlo implementation of Bayesian neural network learning.

4.4.3 Tests on the forensic glass data

The forensic glass data was used by Ripley (1994a, 1994b) to test several nonlinear classifiers, including various neural network models. The task is to determine the origin of a fragment of glass found at the scene of a crime, based on measurements of refractive index and of chemical composition (percent by weight of oxides of Na, Mg, Al, Si, K, Ca, Ba, and Fe). The original data set of 214 cases was collected by B. German.[18] Ripley discarded the cases of headlamp glass and randomly divided the remainder into a training set of 89 cases and a test set of 96 cases, which he has kindly made available. The possible classifications in Ripley's data and the

---

[18] This dataset is available via the World Wide Web from the UC Irving Repository of Machine Learning Databases, under the name "glass". The repository is located at URL http://www.ics.uci.edu/~mlearn/MLRepository.html

number of occurrences of each in the training and test sets are as follows: float-processed window glass (30 train, 40 test), non-float-processed window glass (39 train, 37 test), vehicle glass (9 train, 8 test), and other (11 train, 11 test).

I normalized the inputs for this problem to have mean zero and standard deviation one across the training set, as did Ripley. In terms of the rationale discussed in Section 4.3.1, normalization is less justifiable for this problem than for the Boston housing data. There is no obvious way in which the original investigators might have used their beliefs about the problem to control the population from which the data was sampled. The set of input attributes available also appears to simply be all those that could easily be measured, not those that the investigators thought might be most relevant. It is therefore difficult to see how normalization can act as a surrogate for input transformations based on expert prior knowledge. Nevertheless, something must be done here, since the inputs as given are very un-normalized, to an extent that appears from my non-expert perspective to be clearly undesirable.

For the network models tested, I used informative priors for the hyperparameters in an attempt to reflect my actual beliefs about the range of plausible values for the weights in various classes. This was done out of concern that vague priors could lead to networks in which the weights into the output units were very large. The softmax model used would then produce conditional distributions for the target given the inputs in which one of the target values has a probability close to one. This corresponds to a belief that, if only we knew enough, the targets would be very predictable, given the inputs. (Note that this situation could not have arisen with the LED display problem of Section 4.3.2, at least when irrelevant inputs are absent or suppressed, because the training sets for that problem contain cases where the relevant inputs are identical but the target is different.)

The possibility that the targets might be perfectly predictable is not completely ruled out by the prior knowledge available. However, it does seem somewhat unlikely — certainly it is at least equally plausible that in many cases the class is ambiguous. If a very vague prior is used for hidden-to-output weights, however, the effect will be to make the posterior probability of perfect predictability for this problem be very high, since when the prior for weight magnitudes extends over a very wide range, large weight magnitudes will dominate the portion of the prior range that is not in strong conflict with the data. This comes about when weights exist that perfectly explain the training data, and which continue to do so as the hidden-to-output weights are scaled up to have arbitrarily large magnitudes. In contrast, scaling down the weights into the outputs by a large factor will result in target distributions that are independent of the inputs, a possibility that will be suppressed in the posterior whenever the

138   Chapter 4. Evaluation of Neural Network Models

training data is predictable to at least some degree. The prior probability for weights of moderate size, resulting in a moderate degree of predictability, will be tiny if the prior is very vague.

The effects of using vague priors for the hyperparameters controlling the input-to-hidden weights are less clear, but I felt it was prudent to avoid extreme values here as well. For one thing, if these weights became very large, the hidden units would effectively compute step functions, and a gradient-based sampling procedure would not be expected to work well.

The network architectures and priors I tried on this problem are described in Figure 4.12. All networks were used in conjunction with the softmax (logistic) model for the targets (Bridle 1989). In accordance with the general philosophy that I advocate, the model that I would choose *a priori* is the most complex one, based on the network with 12 hidden units, using an ARD prior. For all the models, I used a Cauchy prior for the hidden-to-output weights, implemented using a 2-level hierarchical prior, with the low level prior being Gamma with $\alpha = 1$ (see Section 4.1). This choice was somewhat arbitrary — I have no strong reason to think that a Gaussian prior for these weights would be worse. Networks with and without ARD were tried, using informative priors, as discussed above, except for the models based on networks without hidden units (these networks cannot represent a decision boundary that perfectly fits all the training data, and so should not have problems with excessively large weights). One network with vaguer priors was tried as well, to see whether this actually made any difference.

For the networks without hidden units, I ran the Markov chain Monte Carlo procedure for 500 super-transitions, with each super-transition composed of 100 pairs of Gibbs sampling updates and hybrid Monte Carlo updates. The hybrid Monte Carlo trajectories were 100 leapfrog steps long, computed using a stepsize adjustment factor of 0.7. The window-based acceptance procedure was used, with a window of five states. The partial gradient method was not used, since the training set is quite small, and hence is presumably not very redundant. The states saved from the last 300 super-transitions were used to make predictions. These runs each took 4.2 hours, but considerably shorter runs would in fact have been adequate.

For the networks with hidden units, I ran the sampling phase for 200 super-transitions, with a super-transition in this case consisting of 50 pairs of Gibbs sampling and hybrid Monte Carlo updates. The trajectories were 1000 leapfrog steps long, with a window of ten states, and were computed using a stepsize adjustment factor of 0.5. I used the states from the last 100 super-transitions to make predictions. These runs took 18.8 hours for the networks with six hidden units, and 28.6 hours for the network with 12 hidden units. Using the states from the last 50 super-transitions out of the first 100 in these runs gives results that are only a bit worse, however.

	Bias-Output		Input-Output		Hidden-Output		Bias-Hidden		Input-Hidden	
	ω	α	ω	α	ω	α	ω	α	ω	α
Net with no hidden units										
Without ARD	100	0.1	100	0.1	–	–	–	–	–	–
With ARD	100	0.1	100	.001:0.5	–	–	–	–	–	–
Net with 6 hidden units										
Without ARD	100	1	100	1	$100H^2$	1:1	100	1	100	1
With ARD	100	1	100	1:2.5	$100H^2$	1:1	100	1	100	1:2.5
With ARD (vaguer)	100	1	100	.001:0.5	$100H^2$	0.1:1	100	0.1	100	.001:0.5
Net with 12 hidden units										
With ARD	100	1	100	1:2.5	$100H^2$	1:1	100	1	100	1:2.5

FIGURE 4.12. Networks and priors tested on the forensic glass data. The priors for the hyperparameters were all of the Gamma form (equation 4.2). Two-level priors were used for some classes of weights in some models. The top-level mean precision (inverse variance) associated with a group of weights is given by ω, and for hidden-to-output weights is scaled according to the number of hidden units (H). The shape parameters for the Gamma distributions are given by α. For two-level priors, two α values are given — the first controls the shape of the prior for the high-level hyperparameter, which has mean ω; the second controls the shape of the priors for the low-level hyperparameters, whose common mean is given by the high-level hyperparameter.

Computation time might therefore have been cut in half, though we would then have less basis for deciding whether the true equilibrium distribution had been reached.

The predictive performance of these networks is shown in Figure 4.13, along with the results that Ripley (1994a) reports for neural networks and other methods. Performance is judged here by three criteria — mis-classification rate, mis-classification rate with the two types of window glass not distinguished, and average log probability assigned to the correct class. The first two criteria are also used by Ripley. The mis-classification rate is the fraction of test cases for which the best guess produced by the model is not correct, the best guess being the class whose predictive probability is the highest. When the two categories of window glass are combined, the predictive probabilities for each are summed for the purpose of determining the best guess. In a forensic application, a guess without any indication of reliability is perhaps not useful. To test the accuracy of the full predictive distribution produced by the models, I report minus the log of the predictive probability of the correct class, averaged over the test cases.[19]

---

[19] For this problem, it may in fact be inappropriate to use predictive probabilities in any of these ways, since such probabilities take no account of other available information. Instead, the likelihoods for the various classes might be reported; these could then be combined with likelihoods derived from other data, together with a suitable prior. One approach would be to convert the predictive probabilities found here to relative likelihoods by dividing each class's probability by its frequency in the training set.

Model or procedure used	Full error rate	Merged error rate	Average neg log prob
From base rates in training set	61%	20%	1.202
Network with no hidden units			
Without ARD	42%	17%	0.937
With ARD	49%	17%	0.916
Network with six hidden units			
Without ARD (two runs)	28%	14%	0.831
	28%	14%	0.777
With ARD (two runs)	26%	14%	0.765
	27%	14%	0.767
With ARD, vaguer priors	33%	18%	0.873
Network with 12 hidden units			
With ARD	25%	14%	0.745
Network with two hidden units			
Max. penalized likelihood	38%	16%	–
Approx. Bayesian method	38%	14%	–
Network with six hidden units			
Max. penalized likelihood	33%	16%	–
Approx. Bayesian method	28%	12%	–
Linear discriminant	41%	22%	–
Nearest neighbor	26%	17%	–
Projection pursuit	40%	19%	–
Classification tree	28%	15%	–
MARS			
Degree=1	37%	17%	–
Degree=2	31%	19%	–

FIGURE 4.13. Results on the forensic glass data. The figures shown are percent mis-classification, percent mis-classification with the two types of window glass merged, and the average of minus the (natural) log probability of the correct class (where available), all over the test set of 96 cases. The first line shows the performance achieved by simply using the base rates for the classes, taken from their frequencies in the training set. The next section gives results of various Bayesian neural network models trained by Markov chain Monte Carlo. The last two sections give results reported by Ripley (1994a), first for neural networks trained with "weight decay" (maximum penalized likelihood) or by an approximate Bayesian method, second for various other statistical procedures.

4.4 Tests of Bayesian models on real data sets

Note that the test set on which these performance figures are based is quite small (96 cases). Ripley (1994a) considers differences of 4% or less in mis-classification rate to not be significant (at the 5% level), a criterion which I will also use in the assessments below. Note, however, that there is also an unquantified degree of variability with respect to the random choice of training set, which is not taken into account by this criterion. One should therefore treat any observed differences as being suggestive only, as with any comparison of methods that is based on a single training set.

For the networks with no hidden units, use of ARD did not appear to produce any benefit. In fact, the error rate on the full classification task is worse with ARD than without ARD, though the ARD model is slightly better in terms of average log probability for the true target. Use of ARD did have a significant effect on the network weights that were found. In the posterior distribution for the ARD model, the weights from two of the inputs (those giving the percent by weight of oxides of Mg and Al) were substantially bigger than the weights from other inputs, by a factor of almost ten, on average. The corresponding differences in weight magnitudes were much less for the non-ARD model.

The runs for networks with six hidden units produced one strange result. In the first run using a non-ARD prior, the distribution for the magnitudes of input-to-hidden weights changed dramatically around super-transition 80 (out of 200). At this point, these weights changed from magnitudes of less than ten to magnitudes in the hundreds; they may still have been slowly growing at the end of the run. I did another run to see whether this behaviour occurred consistently, and found that in the second run these weights stayed small (magnitudes around ten or less) for the duration. These weights also remained small in two runs using ARD priors. It is possible that the same change seen in the first non-ARD run would have occurred in the second non-ARD run if it had continued for longer, however. It is possible also that the ARD runs might have converged eventually to a distribution in which these weights were large, though it is also plausible that the use of an ARD prior for these weights would change the behaviour.

As shown in Figure 4.13, for the networks with six hidden units, the observed performance of the ARD models was slightly better than that of the non-ARD models, but the differences are not significant, except perhaps with respect to the poor value for average log probability seen with the non-ARD network with large input-to-hidden weights. Use of ARD did appear to have a significant effect of the magnitudes of the weights from different inputs; these magnitudes were more spread out in the ARD runs than in the second non-ARD run. It is difficult to interpret the results, however, since the variation in weight magnitudes between different inputs for a single network is less than the posterior variation in the overall magnitudes for

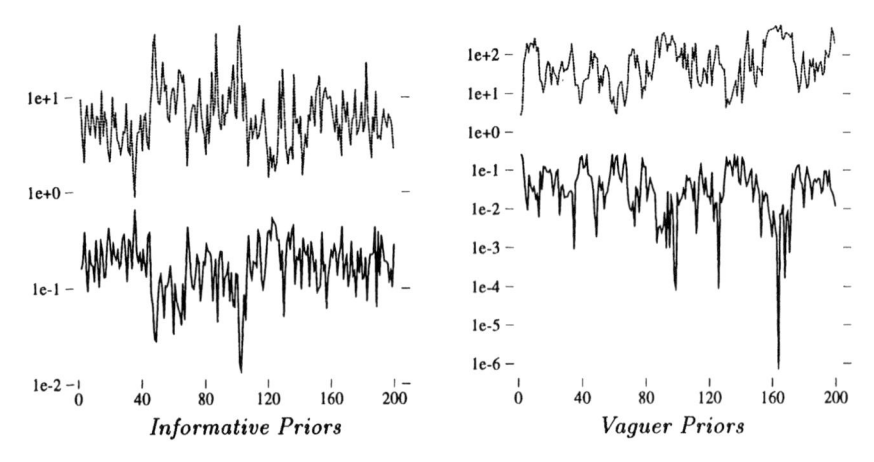

142   Chapter 4. Evaluation of Neural Network Models

FIGURE 4.14. Effect of vague priors in the forensic glass problem. The graphs show the progress of two quantities during the course of simulation runs for networks with six hidden units, using ARD priors. On the left is the run for the model with informative (fairly narrow) priors; on the right is the run for the same model with vaguer priors. The solid line plots the average entropy of the conditional distributions for targets in test cases, as defined by the network parameters from each state (note that this is not the same as the entropy of the predictive distribution, in which we integrate over the posterior). The dotted line plots the square root of the average magnitude of hidden-to-output weights.

input-to-hidden weights. There are also direct connections from inputs to outputs, making it difficult to tell what the total effect of each input is.

One run was done for an ARD model with vaguer priors. In the resulting posterior distribution for network parameters, the hidden-to-output weights had large magnitudes, and the conditional distributions for targets had low entropy, as expected on the basis of the previous discussion. The input-to-hidden weights also had large magnitudes. The effects of using vague priors are illustrated in Figure 4.14; note how the entropy tends to be less when the hidden-to-output weights are large.

Finally, I did a run using a network with 12 hidden units, with an ARD prior. As seen in Figure 4.13, the performance of this model was the best of any tested here, though not all the differences are statistically significant. The results for the ARD network with 12 hidden units and for the ARD networks with 6 hidden units are also not significantly different from that of the network with six hidden units that Ripley (1994a) trained with an approximate Bayesian method based on Gaussian approximations to several modes. All the Bayesian network models trained by Markov chain Monte Carlo (except the one with overly-vague priors) performed significantly better than the other networks trained by Ripley. Of the other statistical techniques that Ripley tried, only the nearest neighbor and classification tree methods performed well. Their observed performance was slightly worse than that of the ARD network with 12 hidden units, but the difference is not statistically significant.

These tests show that vague priors should not be used recklessly. Care in this respect seems to be especially necessary for classification models. The different results obtained from the two runs of the non-ARD model show that one should also not recklessly assume that apparent convergence of a Markov chain sampler is real — clearly, at least one of these two runs did not sample adequately from the true equilibrium distribution. Use of simulated annealing, as in my previous work (Neal 1992b), might help in this respect, though there will still be no guarantees. It would also be interesting to apply a "tempering" method (Marinari and Parisi 1992, Geyer and Thompson 1995, Neal, in press), in order to sample efficiently in cases where the posterior distribution has widely separated modes, which is one possible explanation for the divergence seen here between the two non-ARD runs.

Chapter 5

Conclusions and Further Work

The preceding three chapters have examined the meaning of Bayesian neural network models, showed how these models can be implemented by Markov chain Monte Carlo methods, and demonstrated that such an implementation can be applied in practice to problems of moderate size, with good results. In this concluding chapter, I will review what has been accomplished in these areas, and describe on-going and potential future work to extend these results, both for neural networks and for other flexible Bayesian models.

5.1 Priors for complex models

One major result of this work is that, when an appropriate prior is used, one need not limit the complexity of a network model based on the amount of training data available. This was shown theoretically in Chapter 2, and demonstrated empirically in Chapter 4. In hindsight, at least, the ability to use complex models on small data sets is simply what would be expected, from a Bayesian viewpoint. Nevertheless, it has not been apparent to previous investigators, perhaps because of the pervasive influence of frequentist methods, for which such limits on complexity can indeed be essential.

With the realization that one need not keep networks small, the way was opened for the examination in Chapter 2 of networks with infinite numbers of hidden units. Only in the infinite network limit does it become possible to analytically derive interesting properties of the prior over functions implied by the prior over network parameters.

I first investigated the properties of priors that converge to Gaussian processes as the number of hidden units goes to infinity. These priors can be characterized by their covariance functions. Priors were developed that produce smooth, Brownian, and fractional Brownian functions. Further theoretical work in this area would be of interest. The arguments relating to fractional Brownian priors that I presented could be made more rigorous, and one could contemplate extensions to "multi-fractals", whose properties are different at different scales. The rate of convergence to the infinite network limit could be quantified. It would also be interesting to apply fractional Brownian models to actual data sets. This is supported by the implementation scheme described in Chapter 3 and Appendix A. I have not tried this yet, however, and some difficulties with convergence might be anticipated with such models.

Although the examination of Gaussian priors revealed much of interest, it also showed that such priors are in some respects disappointing. Infinite networks drawn from these priors do not have hidden units that represent "hidden features" of the input. The ability to find such hidden features is often seen an interesting aspect of neural network learning. With Gaussian priors, we also do not see any fundamentally new phenomena when we go to networks with more than one hidden layer — we just get another Gaussian process, albeit with a different covariance function.

Furthermore, for problems where we do feel that these Gaussian process models are appropriate, it may well be that a direct implementation of a Bayesian model based on a Gaussian process would work better in practice than a Bayesian network model that converges to a similar Gaussian process for a large number of hidden units. This possibility, mentioned in Chapter 2, has recently been pursued by Williams and Rasmussen (1996). For a fixed covariance function, Bayesian inference with this model — ie, formation of a predictive distribution for a test case give data on n training cases — can be accomplished using standard matrix operations in $O(n^3)$ time, which is tolerable for training sets containing up to at least several hundred cases. The well-known methods of smoothing splines and kriging are equivalent to certain Bayesian models of this type (Wahba 1990), but for reasons that are mysterious to me, such Gaussian process models have apparently received little consideration for problems with more than two or three dimensions.

In the Gaussian process models used by Williams and Rasmussen (1996), the covariance function is determined by hyperparameters that are analogous to those used in the network models of this book. In light of the theoretical convergence results of Chapter 2, one would expect such a model to perform similarly to a Bayesian network model with one large hidden layer and Gaussian priors. (The models of Williams and Rasmussen do not use exactly the same form of covariance function as that for such a network model, but the covariance functions appear to have similar properties.) In preliminary evaluations, the performance of these two methods was indeed found to be very similar, with both performing better than several competing methods (Rasmussen 1996).

The limitations of Gaussian priors (or any prior with finite variance), and the fact that the models they define may be better handled by a non-network implementation, motivate interest in priors that converge to non-Gaussian stable distributions. A basic convergence result for these priors was derived in Chapter 2, but much work remains to be done in characterizing their properties theoretically, as could perhaps be done using some analogue of the covariance function used to characterize Gaussian processes. Future research could also look at an alternative implementation scheme for such priors based on their representation as Poisson processes (see Section 2.2.1). In such an implementation, the finite number of hidden units available would not come from a finite approximation to the limiting distribution, but would instead be those with the largest hidden-to-output weights from the true limiting distribution. This scheme might make more efficient use of the available hidden units, since resources would not be wasted on units with small weights (and hence little influence). It might also allow one to estimate how much the results could differ from those that would be obtained using the true infinite network.

Some preliminary results concerning priors for networks with more than one hidden layer were reported in Chapter 2, and a network with two hidden layers was found to perform well in the tests on the Boston housing data in Chapter 4. I believe that further work on priors for such networks might produce insights of practical importance. Work on networks with an infinite number of hidden layers would be of at least theoretical interest, in that it would test how far one can push the idea that limiting the complexity of the model is unnecessary.

The theoretical examination of priors in Chapter 2 was supplemented by visual examination of functions drawn from these priors. People have of course looked at samples from priors before. Nevertheless, I believe that this technique is not as widely used as it deserves to be. I hope that my use of it here has demonstrated its utility in developing an intuitive understanding of complex Bayesian models.

5.2 Hierarchical Models — ARD and beyond

Another aspect of prior specification emphasized in this work is the use of hierarchical models whose hyperparameters can adopt values that are appropriate given the characteristics of the data at hand.

One example of this approach is the Automatic Relevance Determination (ARD) model, which is meant to allow the data to determine which inputs should influence the predictions. The tests done on synthetic data in Chapter 4 showed that use of ARD resulted in the suppression of inputs that were unrelated to the prediction task, as well as those that were related, but were superseded by other inputs. The ARD method was also used for the tests on real data, with the result that some inputs were suppressed relative to others, but here the correct behaviour is of course unknown. Further experiments would be required to properly assess the effect of ARD on predictive performance for the real data sets.

In more recent work, I have extended the idea of Automatic Relevance Determination to produce hierarchical models that can determine an appropriate additive structure for a regression function. In an additive regression model (Hastie and Tibshirani 1990), a real-valued target, y, might be modeled as follows, in terms of inputs x_1, x_2, and x_3:

$$y \;=\; g_1(x_1) \;+\; g_2(x_2) \;+\; g_3(x_3) \;+\; \epsilon \qquad (5.1)$$

where g_1, g_2, and g_3 are unknown functions, and ϵ represents Gaussian noise. This form is more general than a linear model, but less general than a model in which y is an arbitrary function of x_1, x_2, and x_3, plus noise. If the above additive form is appropriate for the actual data, using it will have advantages over an unrestricted regression model, both in terms of predictive performance, and in terms of interpretability.

However, just as we will often not be sure which inputs are relevant for predicting a target, y, we will also often be unsure whether an unrestricted or an additive model is more appropriate — or, indeed, whether some intermediate model might be best, such as one in which y is modeled as a function of x_1 and x_2 plus a function of x_3. One could try to somehow identify the "true" model from among the various possibilities, but for many problems, our actual belief will be that none of the additive models can be exactly true (ie, that there is at least some small degree of interaction between all the variables). In such situations, it makes sense to instead use a single hierarchical model that can produce a variety of approximately additive models, as well as an unrestricted model, depending on the settings of its hyperparameters. The data will then be able to select an appropriate form for the regression function, or perhaps several forms, with certain posterior probabilities.

Using a single hierarchical model may also be computationally easier than computing the posterior probabilities of several models with varying degrees of additivity. To calculate posterior model probabilities, one must compute the prior probability of the training data under each model, which is often quite difficult (Neal 1993b, Sections 2.3 and 6.2).

5.2 Hierarchical Models — ARD and beyond

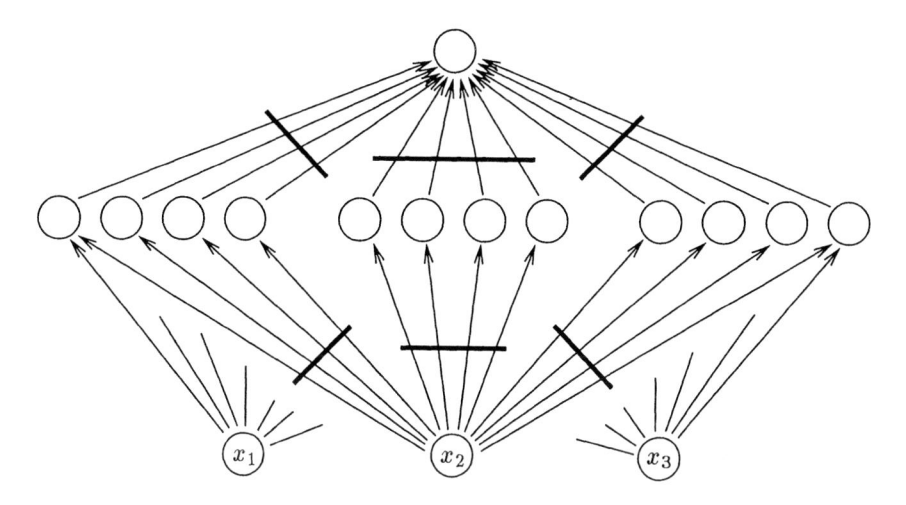

FIGURE 5.1. A hierarchical network model capable of finding additive structure. The network takes three inputs, x_1, x_2, x_3, and computes a single output, which gives the mean of the target value, y. This output is computed using three groups of hidden units, each of which has its own set of hyperparameters, controlling the scale of its contribution to the output, and the degree to which the group pays attention to each of the three inputs. These hyperparameters are represented by heavy lines crossing the connections whose weights they control. The connections out of x_1 and x_3 (and the associated hyperparameters) have been suppressed for clarity.

Figure 5.1 shows a hierarchical model of this sort based on a multilayer perceptron network. This model is essentially three ARD models joined together at the output. Each of these sub-models has its own set of hyperparameters that control the magnitudes of weights on connections into and out of its own group of hidden units. The functions computed by each sub-model (ie, the weighted sum of the values of the hidden units in each group) are added together to produce the network output, which is interpreted as the mean of the conditional distribution for the target, y.

If an additive decomposition of the regression function is in fact appropriate, we would hope that the posterior distribution for this model will be concentrated on sets of values for the hyperparameters in which each group of hidden units pays attention to exactly one of the three inputs, so that the three groups will compute the three functions, $g_1(x)$, $g_2(x)$, and $g_3(x)$, of equation (5.1). If, on the other hand, the three variables interact, we would hope that at least one of the groups of hidden units will end up with hyperparameter values that allow it to look at all three inputs. If there is no additive structure at all, we might expect the contribution to the output of all but one of the groups of hidden units to be suppressed, by means of the hyperparameters controlling their hidden-to-output weights.

Other structures, such as the intermediate model mentioned above, should also be possible. With any of these structures, inputs whose effects have been largely suppressed may still have a slight influence, as will be desirable when we do not believe that the true function has an exactly additive form.

From preliminary experiments, it appears that this scheme does indeed operate as desired — finding additive structure when it is present, and not finding it when it is not present. Furthermore, the predictive performance of a network with the multiple groups of hidden units shown in Figure 5.1 can be better than that of a simple ARD model when the function does indeed have additive structure. These models push the limits of the present Markov chain implementation, as they require that the Markov chain explore a complex space of possible hyperparameter values. The problem of random walks discussed in Chapter 3 is of significance here, as the hyperparameters are presently updated using Gibbs sampling, which does not suppress random walk behaviour. Exploration may also be inhibited by local modes in the posterior distribution over hyperparameters.

Hierarchical models with multiple groups of hidden units may be useful in other contexts as well. With appropriate sets of connections and hyperparameters, models can be defined that have the potential to produce functions with more than one scale of variation, to select between a Gaussian or non-Gaussian prior for hidden-to-output weights, or to select between a one-layer and a two-layer network. As with ARD models and the model of Figure 5.1, the actual result of applying such models might be a network that only approximately satisfies some restricted form, or a posterior distribution over several network structures with substantially different characteristics.

A final cautionary note regarding hierarchical models: The tests in Chapter 4 indicate that care is required when using vague priors for hyperparameters. Such priors are a convenience, since they allow one to avoid thinking about what the exactly appropriate prior would be; they also work well in some contexts. The results on the forensic glass data show that vague priors can sometimes lead to problems, however, especially with classification models. One might also expect to see problems when using very vague priors with the elaborate hierarchical models discussed above.

5.3 Implementation using hybrid Monte Carlo

Another major theme of this work is the use of a Markov chain Monte Carlo implementation based on the hybrid Monte Carlo algorithm of Duane, Kennedy, Pendleton, and Roweth (1987). I demonstrated in Chapter 3 that hybrid Monte Carlo can be many times faster at sampling the posterior distribution for network weights than simpler forms of the Metropolis algorithm; other methods, such as Gibbs sampling, cannot be applied to this problem at all. Without hybrid Monte Carlo, the Markov chain Monte Carlo approach would not be feasible for any but the smallest networks.

The utility of the hybrid Monte Carlo algorithm extends beyond the neural network field. Although Gibbs sampling and simple forms of the Metropolis algorithm are adequate for many problems of Bayesian inference, I believe that hybrid Monte Carlo can solve many such problems faster than the methods presently used, and will permit the use of complex models for which the computations have hitherto been infeasible. One recent use is for the Gaussian process models of Williams and Rasmussen (1996), discussed in Section 5.1, for which a form of hybrid Monte Carlo is used to sample from the posterior distribution of the hyperparameters controlling the covariance function.

Although the implementation I have described in this thesis is the result of several design iterations, there is no reason to think that it is optimal. The time required for the tests in Chapter 4 shows that improvement in this respect is quite important. Many implementation schemes differing in detail could be investigated. For example, the leapfrog stepsizes could be chosen differently, the hyperparameters could be updated by hybrid Monte Carlo rather than Gibbs sampling, a different parameterization of the weights or the hyperparameters could be used, and the manual methods used to choose a good trajectory length could be improved. Three variants of the basic hybrid Monte Carlo method, using "partial gradients", "windows", and "persistence", were investigated in Chapter 3, and found to give some improvement, especially when used together. Other variants remain to be tried, including those based on discretizations of the dynamics accurate to higher order than the leapfrog method, and possible variants that exploit the (somewhat limited) ability to quickly recalculate the output of a network when a single weight changes (if intermediate results are stored). Finally, one could try applying methods for escaping local modes such as simulated tempering (Marinari and Parisi 1992, Geyer and Thompson 1995) and tempered transitions (Neal, in press).

A topic that was only touched on in Chapter 3 is the use of a Markov chain Monte Carlo implementation to evaluate the accuracy of other implementations, such as those based on Gaussian approximations. It would be most useful if one could use the Monte Carlo implementation to uncover some interesting class of easily-identifiable situations where the Gaussian approximation can be relied upon. This may be too much to hope for, however. Another approach would be to develop procedures whereby Markov chain Monte Carlo methods could be applied to a subset of the training data, at lower computational cost than a full Monte Carlo run, and the results used to assess whether the Gaussian approximation would be adequate when applied to the full data set. On the other hand, it is possible that use of the Markov chain Monte Carlo implementation will in the end prove best in all or most circumstances, once the effort of verifying the validity of Gaussian or other approximations is taken into account.

5.4 Evaluating performance on realistic problems

In Chapter 4, I tested various neural network models on real and synthetic data sets. The main conclusion from these tests is that Bayesian learning implemented using hybrid Monte Carlo can be effectively applied to real problems of moderate size (with tens of inputs, and hundreds of training cases). On one data set (the Boston housing data), the predictive performance obtained using the Bayesian methodology was substantially better than that previously obtained using other methods; on another data set (the forensic glass data), performance was as good as any obtained with other methods. Approximately a day of computation was required to train the networks on these real data sets. This time is large compared to that required by standard methods, but small enough that use of this implementation of Bayesian learning would be practical in many contexts.

Results on only two real data sets are of course not sufficient to support any sweeping claims regarding the superiority of Bayesian learning. More evaluations, on more data sets, in comparison with the best alternative methods, would be required before any conclusions could be drawn that might be accepted by advocates of the methods found to be inferior. It is an unfortunate fact that although performance on real data — or better, on real problems, with real-world context — is the ultimate ground on which learning methods must be judged, fair and comprehensive tests of performance on real problems are quite difficult, and quite laborious, to perform. A group of us at the University of Toronto, led by Geoffrey Hinton, are currently working on the DELVE project, building a large collection of data sets, and an environment that facilitates using this data to make valid comparisons of learning methods on realistic tasks.[1] We hope that we, and other researchers, will soon be able to obtain more definitive evidence regarding the merits of Bayesian learning for neural networks and of other modern methods for solving nonparametric regression and classification tasks.

---

[1]For the latest information on the DELVE project, visit our Web page, at URL http://www.cs.utoronto.ca/neuron/delve/

Appendix A
Details of the Implementation

This appendix contains mathematical details regarding the Bayesian neural network implementation described in Chapter 3, and used for the evaluations in Chapter 4. Some features of this implementation are not discussed in these earlier chapters, but are described here for completeness.

Due to the variety of network architectures accommodated, it is necessary here to use a notation that is more systematic, albeit more cumbersome, than that which is used elsewhere. This notation is summarized on the next page.

A.1 Specifications

This section defines the class of network models that are implemented by the software, and explains how they are specified, in an abstract way. (For the detailed syntax of network specifications, and other non-mathematical details, see the documentation that comes with the software.)

A.1.1 Network architecture

The multilayer perceptron networks that this implementation supports consist of a layer of input units, zero or more hidden layers with tanh activation function, and a layer of output units. Units in each hidden layer are connected to units in the preceding hidden layer and to units in the input

Values associated with units

v_i^I	Value of ith input unit, before the offset is added
v_i^ℓ	Value of ith hidden unit in layer ℓ, before the offset is added
v_i^O	Value of ith output unit
u_i^ℓ	Value of the input to the ith hidden unit

Parameters of the network

t_i^I	Offset for ith input unit
t_i^ℓ	Offset for ith hidden unit in layer ℓ
b_i^ℓ	Bias for ith unit in hidden layer ℓ
b_i^O	Bias for ith output unit
$w_{i,j}^{I,O}$	Weight from ith input unit to jth output unit
$w_{i,j}^{I,\ell}$	Weight from ith input unit to jth unit in hidden layer ℓ
$w_{i,j}^{\ell-1,\ell}$	Weight from ith unit in hidden layer $\ell-1$ to jth unit in hidden layer ℓ
$w_{i,j}^{\ell,O}$	Weight from ith unit in hidden layer ℓ to the jth output unit

Hyperparameters defining priors for parameters

σ_t^I	Common sigma for offsets of input units
σ_t^ℓ	Common sigma for offsets of units in hidden layer ℓ
σ_b^ℓ	Common sigma for biases of units in hidden layer ℓ
σ_b^O	Common sigma for biases of output units
$\sigma_w^{I,O}$	Common sigma for weights from input units to output units
$\sigma_w^{I,\ell}$	Common sigma for weights from input units to units in hidden layer ℓ
$\sigma_w^{\ell-1,\ell}$	Common sigma for weights from units in hidden layer $\ell-1$ to units in hidden layer ℓ
$\sigma_w^{\ell,O}$	Common sigma for weights from units in hidden layer ℓ to output units
$\sigma_{w,i}^{I,O}$	Sigma for weights from input unit i to output units
$\sigma_{w,i}^{I,\ell}$	Sigma for weights from input unit i to units in hidden layer ℓ
$\sigma_{w,i}^{\ell-1,\ell}$	Sigma for weights from unit i in hidden layer $\ell-1$ to units in hidden layer ℓ
$\sigma_{w,i}^{\ell,O}$	Sigma for weights from unit i in hidden layer ℓ to output units
$\sigma_{a,i}^O$	Sigma adjustment for weights and biases into output unit i
$\sigma_{a,i}^\ell$	Sigma adjustment for weights and biases into unit i in hidden layer ℓ

layer. Units in the output layer are connected to units in the hidden layers and to units in the input layer. Each of these connections has an associated weight, used to form a weighted sum of inputs to a unit along incoming connections. Each unit in the hidden and output layers has a bias, which is added to this weighted sum of inputs. Each unit in the input and hidden layers has an offset, which is added to its output. Any of these sets of parameters (associated with a particular layer, or pair of layers) may be missing in any particular network, producing the same effect as if their values were zero.

The following formulas define the outputs, v_i^O, of a network for given values of the inputs, v_i^I. Note that the interpretation of these outputs is determined by the data model, described next.

$$u_i^\ell = b_i^\ell + \sum_k w_{k,i}^{I,\ell}(v_k^I + t_k^I) + \sum_k w_{k,i}^{\ell-1,\ell}(v_k^{\ell-1} + t_k^{\ell-1}) \quad (A.1)$$

$$v_i^\ell = \tanh(u_i^\ell) \quad (A.2)$$

$$v_i^O = b_i^O + \sum_k w_{k,i}^{I,O}(v_k^I + t_k^I) + \sum_\ell \sum_k w_{k,i}^{\ell,O}(v_k^\ell + t_k^\ell) \quad (A.3)$$

Here, and subsequently, the summations are over all units in the appropriate layer, or over all hidden layers (for ℓ). The number of layers and the numbers of units in each layer are part of the architecture specification, but these numbers are not given symbols here. The term in the equation for u_i^ℓ involving layer $\ell-1$ is omitted for the first hidden layer.

A.1.2 Data models

Networks are normally used to define models for the conditional distribution of a set of "target" values given a set of "input" values. There are three sorts of models, corresponding to three sorts of targets — real-valued targets (a "regression" model), binary-valued targets (a "logistic regression" model), and "class" targets taking on values from a (small) finite set (a generalized logistic regression, or "softmax" model). For regression and logistic regression models, the number of target values is equal to the number of network outputs. For the softmax model, there is only one target, with the number of possible values for this target being equal to the number of network outputs.

The distributions for real-valued targets, y_j, in a case with inputs v_i^I may be modeled by independent Gaussian distributions with means given by the corresponding network outputs, and with standard deviations given by the hyperparameters σ_j — the "noise levels" for the targets. The probability density for a target given the associated inputs and the network parameters

156   A. Details of the Implementation

is then

$$P(y_j \mid \text{inputs, parameters}) \;=\; \frac{1}{\sqrt{2\pi}\sigma_j}\exp\bigl(-(y_j - v_j^O)^2/2\sigma_j^2\bigr) \quad (A.4)$$

Alternatively, each case, c, may have its own set of standard deviations, $\sigma_{j,c}$, with the corresponding precisions, $\tau_{j,c} = \sigma_{j,c}^{-2}$, being given Gamma distributions with means of τ_j and shape parameter α_2 (called this for reasons that will become clear later):

$$P(\tau_{j,c} \mid \tau_j) \;=\; \frac{(\alpha_2/2\tau_j)^{\alpha_2/2}}{\Gamma(\alpha_2/2)}\,\tau_{j,c}^{\alpha_2/2 - 1}\exp\bigl(-\tau_{j,c}\alpha_2/2\tau_j\bigr) \quad (A.5)$$

The previous case corresponds to the degenerate Gamma distribution with $\alpha_2 = \infty$. Otherwise, integrating out $\tau_{j,c}$ gives a t-distribution for the target with α_2 "degrees of freedom":

$$P(y_j \mid \text{inputs, parameters})$$
$$=\; \frac{\Gamma((\alpha_2+1)/2)}{\Gamma(\alpha_2/2)\sqrt{\pi\alpha_2}\sigma_j}\bigl[1 + (y_j - v_j^O)^2/\alpha_2\sigma_j^2\bigr]^{-(\alpha_2+1)/2} \quad (A.6)$$

For a logistic regression model, the probability that a binary-valued target, y_j, has the value 1 is given by

$$P(y_j = 1 \mid \text{inputs, parameters}) \;=\; \bigl[1 + \exp(-v_j^O)\bigr]^{-1} \quad (A.7)$$

For a softmax model, the probability that a class target, y, has the value j is given by

$$P(y = j \mid \text{inputs, parameters}) \;=\; \exp(v_j^O)\Big/\sum_k \exp(v_k^O) \quad (A.8)$$

A.1.3 Prior distributions for parameters and hyperparameters

The prior distributions for the parameters of a network are defined in terms of hyperparameters. Conceptually, this implementation provides for one hyperparameter for every parameter, but these lowest-level hyperparameters are not explicitly represented. Mid-level hyperparameters control the distribution of a group of low-level hyperparameters that are all of one type and all associated with the same source unit. High-level (or "common") hyperparameters control the distribution of the mid-level hyperparameters, or of the low-level hyperparameters for parameter types with no mid-level hyperparameters. The same three-level scheme is used for noise levels in regression models.

These hyperparameters are represented in terms of "sigma" values, σ, but their distributions are specified in terms of the corresponding "precisions", $\tau = \sigma^{-2}$, which are given Gamma distributions. The top-level mean is given by a "width" value associated with the parameter type. The shape parameters of the Gamma distributions are determined by "alpha" values associated with each type of parameter. An alpha value of infinity concentrates the entire distribution on the mean, effectively removing one level from the hierarchy. The sigma for a weight may also be multiplied by an "adjustment" value that is associated with the destination unit.

This gives the following generic scheme for the priors for weights:

$$P(\tau_w) = \frac{(\alpha_{w,0}/2\omega_w)^{\alpha_{w,0}/2}}{\Gamma(\alpha_{w,0}/2)} \tau_w^{\alpha_{w,0}/2-1} \exp\left(-\tau_w \alpha_{w,0}/2\omega_w\right) \quad (A.9)$$

$$P(\tau_{w,i} \mid \tau_w) = \frac{(\alpha_{w,1}/2\tau_w)^{\alpha_{w,1}/2}}{\Gamma(\alpha_{w,1}/2)} \tau_{w,i}^{\alpha_{w,1}/2-1} \exp\left(-\tau_{w,i} \alpha_{w,1}/2\tau_w\right) \quad (A.10)$$

$$P(\tau_{a,j}) = \frac{(\alpha_a/2)^{\alpha_a/2}}{\Gamma(\alpha_a/2)} \tau_{a,j}^{\alpha_a/2-1} \exp\left(-\tau_{a,j} \alpha_a/2\right) \quad (A.11)$$

For weights from input units to output units, for example, τ_w will equal $\tau_w^{I,O} = [\sigma_w^{I,O}]^{-2}$, and similarly for $\tau_{w,i}$, while $\tau_{a,j}$ will equal $[\sigma_{a,i}^{O}]^{-2}$. The top-level precision value, ω_w, is derived from the "width" value specified for this type of weight. The positive (possibly infinite) values $\alpha_{w,0}$ and $\alpha_{w,1}$ are also part of the prior specification for input to output weights, while α_a is a specification associated with the output units (note that in this case the "width" value is fixed at one, as freedom to set it would be redundant).

The distribution for a weight from unit i of one layer to unit j of another layer may be Gaussian with mean zero and standard deviation given by $\sigma_{w,i}\sigma_{a,j} = [\tau_{w,i}\tau_{a,j}]^{-1/2}$. That is:

$$P(w_{i,j} \mid \sigma_{w,i}, \sigma_{a,j}) = \frac{1}{\sqrt{2\pi}\sigma_{w,i}\sigma_{a,j}} \exp\left(-w_{i,j}^2/2\sigma_{w,i}^2\sigma_{a,j}^2\right) \quad (A.12)$$

(Here, $w_{i,j}$ represents, for example, $w_{i,j}^{I,O}$, in which case $\sigma_{w,i}$ represents $\sigma_{w,i}^{I,O}$ and $\sigma_{a,j}$ represents $\sigma_{a,j}^{O}$.)

Alternatively, each individual weight may have its own "sigma", with the corresponding precision having a Gamma distribution with mean $\tau_{w,i}\tau_{a,j}$ and shape parameter given by $\alpha_{w,2}$. The previous case corresponds to the degenerate distribution with $\alpha_{w,2} = \infty$. Otherwise, we can integrate over the individual precisions and obtain t-distributions for each weight:

$$P(w_{i,j} \mid \sigma_{w,i}, \sigma_{a,j}) \quad (A.13)$$
$$= \frac{\Gamma((\alpha_{w,2}+1)/2)}{\Gamma(\alpha_{w,2}/2)\sqrt{\pi\alpha_{w,2}}\,\sigma_{w,i}\sigma_{a,j}} \left[1 + w_{i,j}^2/\alpha_{w,2}\sigma_{w,i}^2\sigma_{a,j}^2\right]^{-(\alpha_{w,2}+1)/2}$$

A. Details of the Implementation

The same scheme is used for biases, except that for them there are no mid-level hyperparameters. We have

$$P(\tau_b) = \frac{(\alpha_{b,0}/2\omega_b)^{\alpha_{b,0}/2}}{\Gamma(\alpha_{b,0}/2)} \tau_b^{\alpha_{b,0}/2-1} \exp\left(-\tau_b \alpha_{b,0}/2\omega_b\right) \quad (A.14)$$

where τ_b might, for example, be $\tau_b^O = [\sigma_b^O]^{-2}$, etc.

The distribution of the biases is then either

$$P(b_i \mid \sigma_b, \sigma_{a,i}) = \frac{1}{\sqrt{2\pi}\sigma_b \sigma_{a,i}} \exp\left(-b_i^2/2\sigma_b^2\sigma_{a,i}^2\right) \quad (A.15)$$

if $\alpha_{b,1} = \infty$, or if not

$$P(b_i \mid \sigma_b, \sigma_{a,i})$$

$$= \frac{\Gamma((\alpha_{b,1}+1)/2)}{\Gamma(\alpha_{b,1}/2)\sqrt{\pi\alpha_{b,1}}\,\sigma_b\sigma_{a,i}} \left[1 + b_i^2/\alpha_{b,1}\sigma_b^2\sigma_{a,i}^2\right]^{-(\alpha_{b,1}+1)/2} \quad (A.16)$$

For the offsets added to input and hidden unit values, there are no mid-level hyperparameters, and neither are "adjustments" used. We have

$$P(\tau_t) = \frac{(\alpha_{t,0}/2\omega_t)^{\alpha_{t,0}/2}}{\Gamma(\alpha_{t,0}/2)} \tau_t^{\alpha_{t,0}/2-1} \exp\left(-\tau_t \alpha_{t,0}/2\omega_t\right) \quad (A.17)$$

where τ_t might, for example, be $\tau_t^I = [\sigma_t^I]^{-2}$, etc.

The distribution of the offsets is then either

$$P(t_i \mid \sigma_t) = \frac{1}{\sqrt{2\pi}\sigma_t} \exp\left(-t_i^2/2\sigma_t^2\right) \quad (A.18)$$

if $\alpha_{t,1} = \infty$, or if not

$$P(t_i \mid \sigma_t) = \frac{\Gamma((\alpha_{t,1}+1)/2)}{\Gamma(\alpha_{t,1}/2)\sqrt{\pi\alpha_{t,1}}\,\sigma_t} \left[1 + t_i^2/\alpha_{t,1}\sigma_t^2\right]^{-(\alpha_{t,1}+1)/2} \quad (A.19)$$

The scheme for noise levels in regression models is also similar, with τ_j, the precision for target j, being specified in terms of an overall precision, τ, as follows:

$$P(\tau) = \frac{(\alpha_0/2\omega)^{\alpha_0/2}}{\Gamma(\alpha_0/2)} \tau^{\alpha_0/2-1} \exp\left(-\tau\alpha_0/2\omega\right) \quad (A.20)$$

$$P(\tau_j \mid \tau) = \frac{(\alpha_1/2\tau)^{\alpha_1/2}}{\Gamma(\alpha_1/2)} \tau_j^{\alpha_1/2-1} \exp\left(-\tau_j\alpha_1/2\tau\right) \quad (A.21)$$

where ω, α_0, and α_1 are parts of the noise specification. A third alpha (α_2) is needed for the final specification of the noise in individual cases, as described in the Section A.1.2.

A.1.4 Scaling of priors

The top-level mean precisions used in the preceding hierarchical priors (the ω values) may simply be taken as specified (actually, what is specified is the corresponding "width", $\omega^{-1/2}$). Alternatively, for connection weights only (not biases and offsets), the ω for values of one type may be scaled automatically, based on the number of source units that feed into each destination unit via connections of this type. This scaling is designed to produced sensible results as the number of source units goes to infinity, while all other specifications remain unchanged.

The theory behind this scaling concerns the convergence of sums of independent random variables to "stable distributions" (Feller 1966, Samorodnitsky and Taqqu 1994), as discussed in Chapter 2. The symmetric stable distributions are characterized by a width parameter and an index, α, in the range (0,2]. If $X_1, \ldots, X_n$ are independent and each has the same symmetric stable distribution of index α, then $(X_1 + \cdots + X_n)/n^{1/\alpha}$ has this same stable distribution as well. The stable distribution with index 2 is the Gaussian. The sums of all random variables with finite variance converge to the Gaussian, along with some others. Typically, random variables whose moments are defined up to but not including α converge to the stable distribution with index α, for $\alpha < 2$.

This leads to the following scaling rules for producing ω based on the specified base precision, ω_0, the number of source units, n, and the relevant α value (see below):

$$\omega = \begin{cases} \omega_0 n & \text{for } \alpha - \infty \\ \omega_0 n \alpha/(\alpha-2) & \text{for } \alpha > 2 \\ \omega_0 n \log n & \text{for } \alpha = 2 \quad \text{(but fudged to } \omega_0 n \text{ if } n < 3) \\ \omega_0 n^{2/\alpha} & \text{for } \alpha < 2 \end{cases} \quad (A.22)$$

Here, α is $\alpha_{w,2}$ if that is finite, and is otherwise $\alpha_{w,1}$. The scheme doesn't really work if both $\alpha_{w,1}$ and $\alpha_{w,2}$ are finite. When $\alpha = 2$, the scaling produces convergence to the Gaussian distribution, but with an unusual scale factor, as the t-distribution with $\alpha = 2$ is in the "non-normal" domain of attraction of the Gaussian distribution.

A.2 Conditional distributions for hyperparameters

Implementation of Gibbs sampling for hyperparameters requires sampling from the conditional distribution for one hyperparameter given the values of the other hyperparameters and of the network parameters. This section describes how this is done.

A.2.1 Lowest-level conditional distributions

The simplest conditional distributions to sample from are those for "sigma" hyperparameters that directly control a set of network parameters. This will be the situation for the lowest-level sigmas, as well as for higher-level sigmas when the lower-level sigmas are tied exactly to this higher-level sigma (i.e. when the "alpha" shape parameter for their distribution is infinite). The situation is analogous for sigma values relating to noise levels in regression models, except that the errors in training case are what is modeled, rather than the network parameters.

In general, we will have some hyperparameter $\tau = \sigma^{-2}$ that has a Gamma prior, with shape parameter we will call α, and with mean ω (which may be a higher-level hyperparameter). The purpose of τ is to specify the precisions for the independent Gaussian distributions of n lower-level quantities, z_i. In this situation, the conditional distribution for τ will be given by the following proportionality:

$$P(\tau \mid \{z_i\}, \ldots) \propto \tau^{\alpha/2-1} \exp(-\tau\alpha/2\omega) \cdot \prod_i \tau^{1/2} \exp(-\tau z_i^2/2) \quad (A.23)$$

$$\propto \tau^{(\alpha+n)/2-1} \exp\left(-\tau(\alpha/\omega + \sum_i z_i^2)/2\right) \quad (A.24)$$

The first factor in equation (A.23) derives from the prior for τ, the remaining factors from the effect of τ on the probabilities of the z_i. The result is a Gamma distribution that can be sampled from by standard methods (Devroye 1986).

When the distributions of the z_i are influenced by "adjustments", $\tau_{a,i}$, the above formula is modified as follows:

$$P(\tau \mid \{z_i\}, \{\tau_{a,i}\}, \ldots)$$
$$\propto \tau^{(\alpha+n)/2-1} \exp\left(-\tau(\alpha/\omega + \sum_i \tau_{a,i} z_i^2)/2\right) \quad (A.25)$$

Gibbs sampling for the adjustments themselves is done in similar fashion, using the weighted sum of squares of parameters influenced by the adjustment, with the weights in this case being the precisions associated with each parameter.

A.2.2 Higher-level conditional distributions

Sampling from the conditional distribution for a sigma hyperparameter that controls a set of lower-level sigmas is more difficult, but can be done in the most interesting cases using rejection sampling. This method is generally adequate, but not completely satisfactory. I plan to replace it with a better scheme soon.

Assume that we wish to sample from the distribution for a precision hyperparameter τ, which has a higher-level Gamma prior specified by α_0 and ω, and which controls the distributions of lower-level hyperparameters, τ_i, that have independent Gamma distributions with shape parameter α_1 and mean τ. The conditional distribution for τ is then given by the following proportionality:

$$P(\tau \mid \{\tau_i\}, \ldots)$$

$$\propto \tau^{\alpha_0/2-1} \exp(-\tau\alpha_0/2\omega) \cdot \prod_i \tau^{-\alpha_1/2} \exp(-\tau_i\alpha_1/2\tau) \quad (A.26)$$

$$\propto \tau^{(\alpha_0-n\alpha_1)/2-1} \exp\left(-\tau\alpha_0/2\omega - (\alpha_1 \sum_i \tau_i)/2\tau\right) \quad (A.27)$$

Defining $\gamma = 1/\tau$, we get:

$$P(\gamma \mid \{\tau_i\}, \ldots)$$

$$\propto \tau^2 P(\tau \mid \{\tau_i\}, \ldots) \quad (A.28)$$

$$\propto \tau^{(\alpha_0-n\alpha_1)/2+1} \exp\left(-\tau\alpha_0/2\omega - (\alpha_1 \sum_i \tau_i)/2\tau\right) \quad (A.29)$$

$$\propto \gamma^{(n\alpha_1-\alpha_0)/2-1} \exp\left(-\gamma(\alpha_1 \sum_i \tau_i)/2\right) \cdot \exp\left(-\alpha_0/2\omega\gamma\right) \quad (A.30)$$

The first part of this has the form of a Gamma distribution for γ, provided $n\alpha_1 > \alpha_0$; the last factor lies between zero and one. If $n\alpha_1 > \alpha_0$, we can therefore obtain a value from the distribution for γ by repeatedly sampling from the Gamma distribution with shape parameter $n\alpha_1 - \alpha_0$ and mean $(n\alpha_1 - \alpha_0)/(\alpha_1 \sum_i \tau_i)$ until the value of γ generated passes an acceptance test, which it does with probability $\exp(-\alpha_0/2\omega\gamma)$. We may hope that the probability of rejection will be reasonably low if α_0 is small, which is typical.

In some contexts, the values τ_i are not explicitly represented, and must themselves be found by sampling using the method of the previous section.

A.3 Calculation of derivatives

To use the hybrid Monte Carlo method, we must be able to calculate the derivatives of the log of the posterior probability density for the parameter values, which are found by summing the derivatives of the log likelihood and of the log of the prior probability density of the parameter values. This section details how this is done.

A.3.1 Derivatives of the log prior density

For fixed values of the explicitly-represented hyperparameters, one can easily obtain the derivatives of the log of the prior probability with respect to the network weights and other parameters. Generically, if $\alpha_{w,2} = \infty$, we get, from equation (A.12), that

$$\frac{\partial}{\partial w_{i,j}} \log P(w_{i,j} \mid \sigma_{w,i}, \sigma_{a,j}) = -\frac{w_{i,j}}{\sigma_{w,i}^2 \sigma_{a,j}^2} \tag{A.31}$$

while otherwise, we get, from equation (A.14), that

$$\frac{\partial}{\partial w_{i,j}} \log P(w_{i,j} \mid \sigma_{w,i}, \sigma_{a,j})$$

$$= -\frac{\alpha_{w,2} + 1}{\alpha_{w,2} \sigma_{w,i}^2 \sigma_{a,j}^2} \frac{w_{i,j}}{\left[1 + w_{i,j}^2 / \alpha_{w,2} \sigma_{w,i}^2 \sigma_{a,j}^2\right]} \tag{A.32}$$

Similar formulas for derivatives with respect to the biases are obtained from equations (A.15) and (A.16) and for derivatives with respect to the offsets from equations (A.18) and (A.19).

A.3.2 Log likelihood derivatives with respect to unit values

The starting point for calculating the derivatives of the log likelihood with respect to the network parameters is to calculate the derivative of the log likelihood due to a particular case with respect to the network outputs. For the regression model with $\alpha_2 = \infty$, we get from equation (A.4) that

$$\frac{\partial}{\partial v_j^O} \log P(y \mid v_j^O) = -\frac{y_j - v_j^O}{\sigma_j^2} \tag{A.33}$$

When α_2 is finite, we get from equation (A.6) that

$$\frac{\partial}{\partial v_j^O} \log P(y \mid v_j^O) = -\frac{\alpha_2 + 1}{\alpha_2 \sigma_j^2} \frac{y_j - v_j^O}{\left[1 + (y_j - v_j^O)^2 / \alpha_2 \sigma_j^2\right]} \tag{A.34}$$

For the model of binary targets given by equation (A.7), we get the following, after some manipulation:

$$\frac{\partial}{\partial v_j^O} \log P(y \mid v_j^O) = y_j - \left[1 + \exp(-v_j^O)\right]^{-1} \tag{A.35}$$

$$= y_j - P(y_j = 1 \mid v_j^O) \tag{A.36}$$

For the many-way "softmax" classification model of equation (A.8), we get the following (where $\delta(y, j)$ is one if $y = j$ and zero otherwise):

$$\frac{\partial}{\partial v_j^O} \log P(y \mid \{v_k^O\}) = \delta(y, j) - \frac{\exp(v_j^O)}{\sum_k \exp(v_k^O)} \tag{A.37}$$

$$= \delta(y,j) - P(y=j \mid \{v_k^O\}) \quad (A.38)$$

Let L be the log likelihood due to a single training case — that is, $L = \log P(y \mid \text{inputs, parameters}) = \log P(y \mid \text{outputs})$. Once the derivatives of L with respect to the output unit values are known, its derivatives with respect to the values of the hidden and input units can be found by the standard backpropagation method. From equations (A.1), (A.2), and (A.3):

$$\frac{\partial L}{\partial v_i^\ell} = \sum_j w_{i,j}^{\ell,O} \frac{\partial L}{\partial v_j^O} + \sum_j w_{i,j}^{\ell,\ell+1} \frac{\partial L}{\partial u_j^{\ell+1}} \quad (A.39)$$

$$\frac{\partial L}{\partial u_i^\ell} = (1 - [v_i^\ell]^2) \frac{\partial L}{\partial v_j^\ell} \quad (A.40)$$

$$\frac{\partial L}{\partial v_i^I} = \sum_j w_{i,j}^{I,O} \frac{\partial L}{\partial v_j^O} + \sum_\ell \sum_j w_{i,j}^{I,\ell} \frac{\partial L}{\partial u_j^\ell} \quad (A.41)$$

In (A.39), the second term is not present when ℓ is the last hidden layer.

A.3.3 Log likelihood derivatives with respect to parameters

The derivatives of L with respect to the network parameters (with explicitly represented noise sigmas fixed) are obtained using the derivatives with respect to unit values and unit inputs found in the previous section, as follows:

$$\frac{\partial L}{\partial b_i^O} = \frac{\partial L}{\partial v_i^O} \quad (A.42)$$

$$\frac{\partial L}{\partial b_i^\ell} = \frac{\partial L}{\partial u_i^\ell} \quad (A.43)$$

$$\frac{\partial L}{\partial t_i^\ell} = \frac{\partial L}{\partial v_i^\ell} \quad (A.44)$$

$$\frac{\partial L}{\partial t_i^I} = \frac{\partial L}{\partial v_i^I} \quad (A.45)$$

$$\frac{\partial L}{\partial w_{i,j}^{\ell,O}} = \frac{\partial L}{\partial v_j^O} (v_i^\ell + t_i^\ell) \quad (A.46)$$

$$\frac{\partial L}{\partial w_{i,j}^{\ell-1,\ell}} = \frac{\partial L}{\partial u_j^\ell} (v_i^{\ell-1} + t_i^{\ell-1}) \quad (A.47)$$

$$\frac{\partial L}{\partial w_{i,j}^{I,\ell}} = \frac{\partial L}{\partial u_j^\ell} (v_i^I + t_i^I) \quad (A.48)$$

$$\frac{\partial L}{\partial w_{i,j}^{I,O}} = \frac{\partial L}{\partial v_j^O} (v_i^I + t_i^I) \quad (A.49)$$

164   A. Details of the Implementation

The derivatives found in this way for each training case are summed over the full training set, and added to the derivatives with respect to the log prior density, to give the derivatives with respect to the log posterior probability density, which control the hybrid Monte Carlo dynamics.

A.4 Heuristic choice of stepsizes

Stepsizes for dynamical trajectory computations and for Metropolis updates are heuristically chosen based on the values of the training inputs and the current values of the hyperparameters. These stepsize choices are made on the assumption that the system is near equilibrium, moving about in an approximately Gaussian hump of the posterior distribution. If the axes of this hump were aligned with the coordinate axes, the optimal stepsize along each axis would be in the vicinity of the standard deviation along that axis. Since the axes of the bowl may not be aligned with the coordinate axes, the actual stepsizes may have to be less than this. On the other hand, the estimates used are in some respects conservative. Any overall adjustment of the stepsizes to account for these factors must be done manually by the user.

Estimates of the posterior standard deviations along the axes are based on estimates of the second derivatives of the log posterior probability density along the axes. These second derivatives are estimated using estimates of the second derivatives of the log likelihood with respect to the values of units in the network.

Letting L be the log likelihood for a single training case, we get the following for real-valued targets, with $\alpha_2 = \infty$, using equation (A.33):

$$-\frac{\partial^2 L}{\partial (v_j^o)^2} = \frac{1}{\sigma_j^2} \qquad (A.50)$$

while for finite α_2, we get from equation (A.34) that

$$-\frac{\partial^2 L}{\partial (v_j^o)^2} = \frac{\alpha_2 + 1}{\alpha_2 \sigma_j^2} \left[\left(1 + \frac{(v_j^o)^2}{\alpha_2 \sigma_j^2}\right)^{-1} + \frac{2(v_j^o)^2}{\alpha_2 \sigma_j^2} \left(1 + \frac{(v_j^o)^2}{\alpha_2 \sigma_j^2}\right)^{-2} \right] \qquad (A.51)$$

This is estimated by its maximum value, which occurs at $v_j^o = 0$:

$$-\frac{\partial^2 L}{\partial (v_j^o)^2} \approx \frac{\alpha_2 + 1}{\alpha_2 \sigma_j^2} \qquad (A.52)$$

For binary-valued targets, equation (A.36) gives

$$-\frac{\partial^2 L}{\partial (v_j^o)^2} = \frac{1}{[1 + \exp(v_j^o)][1 + \exp(-v_j^o)]} \approx \frac{1}{4} \qquad (A.53)$$

Again, the estimate is based on the maximum possible value, which occurs when $v_j^O = 0$.

We get a similar estimate for a class target, using equation (A.38):

$$-\frac{\partial^2 L}{\partial (v_j^O)^2} = \frac{\exp(v_j^O)}{\sum_k \exp(v_k^O)} \left[1 - \frac{\exp(v_j^O)}{\sum_k \exp(v_k^O)} \right] \approx \frac{1}{4} \qquad (A.54)$$

These estimates for the second derivatives of L with respect to the outputs are propagated backward to give estimates for the second derivatives of L with respect to the values of hidden and input units.

When doing this backward propagation, we need an estimate of the second derivative of L with respect to the summed input to a tanh hidden unit, given its second derivative with respect to the unit's output. Letting the hidden unit output be $v = \tanh(u)$, we have

$$\frac{d^2 L}{du^2} = \frac{d}{du}\left[(1-v^2)\frac{dL}{dv}\right] \qquad (A.55)$$

$$= (1-v^2)^2 \frac{d^2 L}{dv^2} - 2v(1-v^2)\frac{dL}{dv} \qquad (A.56)$$

$$\approx (1-v^2)^2 \frac{d^2 L}{dv^2} \approx \frac{d^2 L}{dv^2} \qquad (A.57)$$

The first approximation assumes that since $2v(1-v^2)(dL/dv)$ may be either positive or negative, its effects will (optimistically) cancel when averaged over the training set. Since v is not known, the second approximation above takes the maximum with respect to v. The end result is that we just ignore the fact that the hidden unit input is passed through the tanh function.

The backward propagation also ignores any interactions between multiple connections from a unit. Since the stepsizes chosen are not allowed to depend on the actual values of the network parameters, the magnitude of each weight is taken to be equal to the corresponding sigma hyperparameter, multiplied by the destination unit adjustment, if present. This gives the following generic estimate:

$$\frac{\partial^2 L}{\partial (v_i^S)^2} \approx \sum_D \sum_j (\sigma_{w,i}^{S,D} \sigma_{a,j}^D)^2 \frac{\partial^2 L}{(v_j^D)^2} \qquad (A.58)$$

Here, S is the source layer, D goes through the various layers receiving connections from S, $\sigma_{w,i}^{S,D}$ is the hyperparameter controlling weights to layer D out of unit i in S, and $\sigma_{a,j}^D$ is the sigma adjustment for unit j in D.

The second derivative of L with respect to a weight, $w_{i,j}^{S,D}$, can be expressed as follows:

$$\frac{\partial^2 L}{\partial (w_{i,j}^{S,D})^2} = (v_i^S)^2 \frac{\partial^2 L}{\partial (v_j^D)^2} \tag{A.59}$$

When the weight is on a connection from an input unit, $v_i^S = v_i^I$ is the ith input for this training case, which is known. If the weight is on a connection from a hidden unit, $(v_i^S)^2$ is assumed to be one, the maximum possible value.

Second derivatives with respect to biases and offsets are simply equal to the second derivatives with respect to the associated unit values.

These heuristic estimates for the second derivatives of L due to each training case with respect to the various network parameters are summed for all cases in the training set. To these are added estimates of the second derivatives of the log prior probability density with respect to each parameter, giving estimates of the second derivatives of the log posterior density.

For the second derivative of the log prior density with respect to weight $w_{i,j}$, we have

$$-\frac{\partial^2}{\partial w_{i,j}^2} \log P(w_{i,j} \mid \sigma_{w,i}, \sigma_{a,j}) = \frac{1}{\sigma_{w,i}^2 \sigma_{a,j}^2} \tag{A.60}$$

if α_2 is infinite, while for finite α_2, we use an estimate analogous to equation (A.52):

$$-\frac{\partial^2}{\partial w_{i,j}^2} \log P(w_{i,j} \mid \sigma_{w,i}, \sigma_{a,j}) \approx \frac{\alpha_2 + 1}{\alpha_2 \sigma_{w,i}^2 \sigma_{a,j}^2} \tag{A.61}$$

Biases and offsets are handled similarly.

Finally, the stepsize used for a parameter is the reciprocal of the square root of minus the estimated second derivative of the log posterior with respect to that parameter.

A.5 Rejection sampling from the prior

In addition to the Monte Carlo implementation based on Markov chain sampling, a simple Monte Carlo procedure using rejection sampling has also been implemented. This procedure is very inefficient; it is intended for use only as a means of checking the correctness of the Markov chain implementations.

The rejection sampling procedure is based on the idea of producing a sample from the posterior by generating networks from the prior, and then accepting some of these networks with a probability proportional to the likelihood (for the given training data) of the generated parameter and

A.5 Rejection sampling from the prior

hyperparameter values. For data models with discrete targets, this idea can be implemented directly, as the likelihood is the probability of the targets in the training set, which can be no more than one. For regression models, the likelihood is the probability density of the targets, which can be greater than one, making its direct use as an acceptance probability invalid. If the noise levels for the targets are fixed, however, the likelihood is bounded, and can be used as the acceptance probability after rescaling. For a Gaussian noise model (equation (A.4)), this is accomplished by simply ignoring the factors of $1/\sqrt{2\pi}\sigma_j$ in the likelihood; the analogous procedure can be used for noise from a t-distribution (equation (A.6)).

When the noise levels are variable hyperparameters, a slightly more elaborate procedure must be used, in which the noise levels are not generated from the prior, but rather from the prior multiplied by a bias factor that gives more weight to higher precisions (lower noise). This bias factor is chosen so that when it is cancelled by a corresponding modification to the acceptance probability, these probabilities end up being no greater than one.

Specifically, the overall noise precision, τ, and the noise precisions for individual targets, the τ_j, are sampled from Gamma distributions obtained by modifying the priors of equations (A.20) and (A.21) as follows:

$$f(\tau) \propto \tau^{nm/2} P(\tau) \qquad (A.62)$$
$$\propto \tau^{(\alpha_0+nm)/2-1} \exp\left(-\tau\alpha_0/2\omega\right) \qquad (A.63)$$
$$f(\tau_j \mid \tau) \propto \tau_j^{n/2} P(\tau_j \mid \tau) \qquad (A.64)$$
$$\propto \tau^{-(\alpha_1+n)/2} \tau_j^{(\alpha_1+n)/2-1} \exp\left(-\tau_j\alpha_1/2\tau\right) \qquad (A.65)$$

Here, n is the number of training cases and m is the number of targets. The resulting joint sampling density is

$$f(\tau, \{\tau_j\}) = f(\tau) \prod_{j=1}^{m} f(\tau_j \mid \tau) \propto P(\tau, \{\tau_j\}) \prod_{j=1}^{m} \tau_j^{n/2} \qquad (A.66)$$

Since this sampling density is biased in relation to the prior by the factor $\prod_{j=1}^{m} \tau_j^{n/2}$, when constructing the acceptance probability we must multiply the likelihood by the inverse of this factor, $\prod_{j=1}^{m} \tau_j^{-n/2} = \prod_{c=1}^{n} \prod_{j=1}^{m} \sigma_j$. This cancels the factors of $1/\sigma_j$ in the target probabilities of equations (A.4) and (A.6), leaving an acceptance probability which is bounded, and can be adjusted to be no more than one by ignoring the remaining constant factors.

Appendix B
Obtaining the software

The implementation of Bayesian learning for neural networks described in Appendix A is available free of charge for research and educational purposes. This implementation is written in C, and currently is designed for use only on Unix systems. It does not require any special graphics or user interface environment. The software also does not use any special Unix facilities, but it is nevertheless likely that various modifications would be required in order for it to run in some other environment, and I cannot undertake to provide assistance with any such conversion.

Potential users should note that this software is intended to support research in Bayesian neural network learning, not as a tool for routine data analysis.

The software is available over the Internet, via my World Wide Web home page, at URL

http://www.cs.utoronto.ca/~radford/

It can also be obtained by anonymous ftp to ftp.cs.utoronto.ca, directory pub/radford. Look in the README file there for further instructions.

Unfortunately, it is difficult to say for how long the above instructions will remain valid. If you encounter difficulties, you should be able to find an up-to-date link at Springer-Verlag's Web page, which is currently located at URL

http://www.springer-ny.com/

Bibliography

ACKLEY, D. H., HINTON, G. E., AND SEJNOWSKI, T. J. (1985) "A learning algorithm for Boltzmann machines", *Cognitive Science*, vol. 9, pp. 147-169.

ANDERSEN, H. C. (1980) "Molecular dynamics simulations at constant pressure and/or temperature", *Journal of Chemical Physics*, vol. 72, pp. 2384-2393.

BALDI, P. AND CHAUVIN, Y. (1991) "Temporal evolution of generalization during learning in linear networks",*Neural Computation*, vol. 3, pp. 589-603.

BARNETT, V. (1982) *Comparative Statistical Inference*, Second Edition, New York: John Wiley.

BERGER, J. O. (1985) *Statistical Decision Theory and Bayesian Analysis*, New York: Springer-Verlag.

BERNARDO, J. M. AND SMITH, A. F. M. (1994) *Bayesian Theory*, New York: John Wiley.

BISHOP, C. M. (1995) *Neural Networks for Pattern Recognition*, Oxford University Press.

BOX, G. E. P. AND TIAO, G. C. (1973) *Bayesian Inference in Statistical Analysis*, New York: John Wiley.

BREIMAN, L., FRIEDMAN, J. H., OLSHEN, R. A., AND STONE, C. J. (1984) *Classification and Regression Trees*, Belmont, California: Wadsworth.

BRIDLE, J. S. (1989) "Probabilistic interpretation of feedforward classification network outputs, with relationships to statistical pattern recognition", in F. Fouglemann-Soulie and J. Héault (editors) *Neuro-computing: Algorithms, Architectures and Applications*, New York: Springer-Verlag.

BUNTINE, W. L. AND WEIGEND, A. S. (1991) "Bayesian back-propagation", *Complex Systems*, vol. 5, pp. 603-643.

CREUTZ, M. AND GOCKSCH, A. (1989) "Higher-order hybrid Monte Carlo algorithms", *Physical Review Letters*, vol. 63, pp. 9-12.

CYBENKO, G. (1989) "Approximation by superpositions of a sigmoidal function", *Mathematics of Control, Signals, and Systems*, vol. 2, pp.303-314.

DUANE, S., KENNEDY, A. D., PENDLETON, B. J., AND ROWETH, D. (1987) "Hybrid Monte Carlo", *Physics Letters B*, vol. 195, pp. 216-222.

DEGROOT, M. H. (1970) *Optimal Statistical Decisions*, New York: McGraw-Hill.

DEVROYE, L.(1986) *Non-uniform Random Variate Generation*, New York: Springer-Verlag.

FALCONER, K. (1990) *Fractal Geometry: Mathematical Foundations and Applications*, Chichester: John Wiley.

FELLER, W. (1966) *An Introduction to Probability Theory and its Applications, Volume II*, New York: John Wiley.

FUNAHASHI, K. (1989) "On the approximate realization of continuous mappings by neural networks", *Neural Networks*, vol. 2, pp. 183-192.

GELFAND, A. E. AND SMITH, A. F. M. (1990) "Sampling-based approaches to calculating marginal densities", *Journal of the American Statistical Association*, vol. 85, pp. 398-409.

GELMAN, A., CARLIN, J. B., STERN, H. S., AND RUBIN, D. B. (1995) *Bayesian Data Analysis*, London: Chapman & Hall.

GEMAN, S., BIENENSTOCK, E., AND DOURSAT, R. (1992) "Neural Networks and the Bias/Variance Dilemma", *Neural Computation*, vol. 4, pp. 1-58.

GEMAN, S. AND GEMAN, D. (1984) "Stochastic relaxation, Gibbs distributions and the Bayesian restoration of images", *IEEE Transactions on Pattern Analysis and Machine Intelligence*, vol. 6, pp. 721-741.

GEYER, C. J. AND THOMPSON, E. A. (1995) "Annealing Markov chain Monte Carlo with applications to ancestral inference", *Journal of the American Statistical Association*, vol. 90, pp. 909-920.

GRENANDER, U. (1981) *Abstract Inference*, New York: John Wiley.

HARRISON, D. AND RUBINFELD, D. L. (1978) "Hedonic housing prices and the demand for clean air", *Journal of Environmental Economics and Management*, vol. 5, pp. 81-102.

HASTIE, T. J. AND TIBSHIRANI, R. J. (1990) *Generalized Additive Models*, London: Chapman & Hall.

HERTZ, J., KROGH, A., AND PALMER, R. G. (1991) *Introduction to the Theory of Neural Computation*, Redwood City, California: Addison-Wesley.

HINTON, G. E. AND VAN CAMP, D. (1993) "Keeping neural networks simple by minimizing the description length of the weights", *Proceedings of the Sixth Annual ACM Conference on Computational Learning Theory, Santa Cruz, 1993*, pp. 5-13.

HORNIK, K., STINCHCOMBE, M., AND WHITE, H. (1989) "Multilayer feedforward networks are universal approximators", *Neural Networks*, vol. 2, pp. 359-366.

HOROWITZ, A. M. (1991) "A generalized guided Monte Carlo algorithm", *Physics Letters B*, vol. 268, pp. 247-252.

JEFFREYS, W. H. AND BERGER, J. O. (1992) "Ockham's razor and Bayesian analysis", *American Scientist*, vol. 80, pp. 64-72. See also the discussion in vol. 80, pp. 212-214.

KENNEDY, A. D. (1990) "The theory of hybrid stochastic algorithms", in P. H. Damgaard, et al. (editors) *Probabilistic Methods in Quantum Field Theory and Quantum Gravity*, New York: Plenum Press.

KIRKPATRICK, S., GELATT, C. D., AND VECCHI, M. P. (1983) "Optimization by simulated annealing", *Science*, vol. 220, pp. 671-680.

LE CUN, Y., BOSER, B., DENKER, J. S., HENDERSON, D., HOWARD, R. E., HUBBARD, W., AND JACKEL, L. D. (1990) "Handwritten digit recognition with a back-propagation network", in D. S. Touretzky (editor) *Advances in Neural Information Processing Systems 2*, pp. 396-404, San Mateo, California: Morgan Kaufmann.

LIU, Y. (1994) "Robust parameter estimation and model selection for neural network regression", in J. D. Cowan, G. Tesuaro, and J. Alspector (editors) *Advances in Neural Information Processing Systems 6*, pp. 192-199. San Mateo, California: Morgan Kaufmann.

MACKAY, D. J. C. (1991) *Bayesian Methods for Adaptive Models*, Ph.D thesis, California Institute of Technology.

MACKAY, D. J. C. (1992a) "Bayesian interpolation", *Neural Computation*, vol. 4, pp. 415-447.

MACKAY, D. J. C. (1992b) "A practical Bayesian framework for backpropagation networks", *Neural Computation*, vol. 4, pp. 448-472.

MACKAY, D. J. C. (1992c) "The evidence framework applied to classification networks", *Neural Computation*, vol. 4, pp. 720-736.

MACKAY, D. J. C. (1994a) "Bayesian non-linear modeling for the energy prediction competition", *ASHRAE Transactions*, vol. 100, pt. 2, pp. 1053-1062.

MACKAY, D. J. C. (1994b) "Hyperparameters: Optimise, or integrate out?", in G. Heidbreder, editor, *Maximum Entropy and Bayesian Methods, Santa Barbara, 1993*, Dordrecht: Kluwer.

MACKENZIE, P. B. (1989) "An improved hybrid Monte Carlo method", *Physics Letters B*, vol. 226, pp. 369-371.

McCULLAGH, P. AND NELDER, J. A. (1983) *Generalized Linear Models*, London: Chapman & Hall.

MARINARI, E. AND PARISI, G. (1992) "Simulated tempering: A new Monte Carlo Scheme", *Europhysics Letters*, vol. 19, pp. 451-458.

METROPOLIS, N., ROSENBLUTH, A. W., ROSENBLUTH, M. N., TELLER, A. H., AND TELLER, E. (1953) "Equation of state calculations by fast computing machines", *Journal of Chemical Physics*, vol. 21, pp. 1087-1092.

NEAL, R. M. (1992a) "Bayesian mixture modeling", in C. R. Smith, G. J. Erickson, and P. O. Neudorfer (editors) *Maximum Entropy and Bayesian Methods: Proceedings of the 11th International Workshop on Maximum Entropy and Bayesian Methods of Statistical Analysis, Seattle, 1991*, Dordrecht: Kluwer Academic Publishers.

NEAL, R. M. (1992b) "Bayesian training of backpropagation networks by the hybrid Monte Carlo method", Technical Report CRG-TR-92-1, Dept. of Computer Science, University of Toronto.

NEAL, R. M. (1993a) "Bayesian learning via stochastic dynamics", in C. L. Giles, S. J. Hanson, and J. D. Cowan (editors), *Advances in Neural Information Processing Systems 5*, pp. 475-482, San Mateo, California: Morgan Kaufmann.

NEAL, R. M. (1993b) "Probabilistic inference using Markov Chain Monte Carlo methods", Technical Report CRG-TR-93-1, Department of Computer Science, University of Toronto. Available in Postscript via the World Wide Web, at URL http://www.cs.utoronto.ca/~radford/

NEAL, R. M. (1994) "An improved acceptance procedure for the hybrid Monte Carlo algorithm", *Journal of Computational Physics*, vol. 111, pp. 194-203.

NEAL, R. M. (in press) "Sampling from multimodal distributions using tempered transitions", to appear in *Statistics and Computing*.

PEITGEN, H.-O. AND SAUPE, D. (editors) (1988) *The Science of Fractal Images*, New York: Springer-Verlag.

PRESS, S. J. (1989) *Bayesian Statistics: Principles, Models, and Applications*, New York: John Wiley.

QUINLAN, R. (1993) "Combining instance-based and model-based learning", *Machine Learning: Proceedings of the Tenth International Conference, Amherst, Massachusetts, 1993*, Morgan Kaufmann.

RASMUSSEN, C. E. (1996) "A practical Monte Carlo implementation of Bayesian learning", in D. S. Touretzky, M. C. Mozer, and M. E. Hasselmo (editors) *Advances in Neural Information Processing Systems 8*, MIT Press.

RIPLEY, B D. (1987) *Stochastic Simulation*, New York: John Wiley.

RIPLEY, B. D. (1981) *Spatial Statistics*, New York: John Wiley.

RIPLEY, B. D. (1994a) "Flexible non-linear approaches to classification", in V. Cherkassky, J. H. Friedman, and H. Wechsler (editors) *From Statistics to Neural Networks: Theory and Pattern Recognition Applications*, pp. 105-126, Springer-Verlag.

RIPLEY, B. D. (1994b) "Neural networks and related methods for classification" (with discussion), *Journal of the Royal Statistical Society B*, vol. 56, pp. 409-456.

RIPLEY, B. D. (1996) *Pattern Recognition and Neural Networks*, Cambridge University Press.

RISSANEN, J. (1986) "Stochastic complexity and modeling", *Annals of Statistics*, vol. 14, pp.1080-1100.

ROBERT, C. P. (1995) *The Bayesian Choice*, New York: Springer-Verlag.

ROSSKY, P. J., DOLL, J. D., AND FRIEDMAN, H. L. (1978) "Brownian dynamics as smart Monte Carlo simulation", *Journal of Chemical Physics*, vol. 69, pp. 4628-4633.

RUMELHART, D. E., HINTON, G. E., AND WILLIAMS, R. J. (1986a) "Learning representations by back-propagating errors, *Nature*, vol. 323, pp. 533-536.

RUMELHART, D. E., HINTON, G. E., AND WILLIAMS, R. J. (1986b) "Learning internal representations by error propagation", in D. E. Rumelhart and J. L. McClelland (editors) *Parallel Distributed Processing: Explorations in the Microstructure of Cognition, Volume 1: Foundations*, Cambridge, Massachusetts: MIT Press.

RUMELHART, D. E., MCCLELLAND, J. L., AND THE PDP RESEARCH GROUP (1986) *Parallel Distributed Processing: Explorations in the Microstructure of Cognition, Volume 1: Foundations*, Cambridge, Massachusetts: MIT Press.

SAMORODNITSKY, G. AND TAQQU, M. S. (1994) *Stable Non-Gaussian Random Processes: Stochastic Models with Infinite Variance*, New York: Chapman & Hall.

SCHMITT, S. A. (1969) *Measuring Uncertainty: An Elementary Introduction to Bayesian Statistics*, Reading, Massachussets: Addison-Wesley.

SMITH, A. F. M. AND ROBERTS, G. O. (1993) "Bayesian computation via the Gibbs sampler and related Markov chain Monte Carlo methods" (with discussion), *Journal of the Royal Statistical Society B*, vol. 55, pp. 3-23 (discussion, pp. 53-102).

STONE, M. (1974) "Cross-validatory choice and assessment of statistical predictions" (with discussion), *Journal of the Royal Statistical Society B*, vol. 36, pp. 111-147.

SZELISKI, R. (1989) *Bayesian Modeling of Uncertainty in Low-level Vision*, Boston: Kluwer.

SZU, H. AND HARTLEY, R. (1987) "Fast simulated annealing", *Physics Letters A*, vol. 122, pp. 157-162.

TIERNEY, L. (1994) "Markov chains for exploring posterior distributions", *Annals of Statistics*, vol. 22, pp. 1701-1762.

THODBERG, H. H. (1996) "A review of Bayesian neural networks with an application to near infrared spectroscopy", *IEEE Transactions on Neural Networks*, vol. 7, pp. 56-72.

TOUSSAINT, D. (1989) "Introduction to algorithms for Monte Carlo simulations and their application to QCD", *Computer Physics Communications*, vol. 56, pp. 69-92.

WAHBA, G. (1990) *Spline Models for Observational Data*, Society for Industrial and Applied Mathematics.

WILLIAMS, C. K. I. AND RASMUSSEN, C. E. (1996) "Gaussian processes for regression", in D. S. Touretzky, M. C. Mozer, and M. E. Hasselmo (editors) *Advances in Neural Information Processing Systems 8*, MIT Press.

WOLPERT, D. H. (1993) "On the use of evidence in neural networks", in C. L. Giles, S. J. Hanson, and J. D. Cowan (editors), *Advances in Neural Information Processing Systems 5*, pp. 539-546, San Mateo, California: Morgan Kaufmann.

YOUNG, A. S. (1977) "A Bayesian approach to prediction using polynomials", *Biometrika*, vol. 64, pp. 309-317.

VAPNIK, V. (1982) *Estimation of Dependencies Based on Empirical Data*, translated by S. Kotz, New York: Springer-Verlag.

Index

0-1 loss, 5

absolute error loss, 5, 104, 106
activation function, 11, 31, 75, 153
additive models, 148–150
adjustment values, 39, 157
alpha values, 101, 157
artificial intelligence, 2, 10
autocorrelations, 24, 81–85, 90, 97, 106
Automatic Relevance Determination (ARD), 15–17, 102, 113–116, 148
   1-level vs. 2-level priors for, 123
   alternative to compare with, 114
   magnitudes of weights when using, 120, 123, 136, 141–142
   prior distributions for, 114–115, 125
   tests on LED display problem, 116–122
   tests on robot arm problem, 122–125

backpropagation, 13, 70, 111, 163

Bayes' Rule, 5
Bayesian learning, *see* Bayesian statistics; learning, Bayesian
Bayesian statistics, 3
   books about, 3
   controversy regarding, 2, 5
bias (for a unit), 11, 155
   prior distribution for, 158
bias-variance tradeoff, 8
Boltzmann distribution, *see* canonical distribution
Boltzmann machine, 25
Boston housing data, 127
   computational performance on, 134
   cross-validation tests on, 132–136
   messy aspects of, 127–129
   neural network models for, 129–132
   predictive performance on, 131, 132, 134–136
   preliminary tests on, 129–132
   Quinlan's results on, 133, 134
Brownian functions, 35–37

candidate state, 26
canonical distribution, 57, 69
    invariance under Hamiltonian dynamics, 59
    over phase space, 57
CART (Classification and Regression Trees), 116, 119
Cauchy distribution, 35, 43, 46
Central Limit Theorem, 32
central region, 36, 52–53
classification models, 12, 14, 31, 150, 155
coin tossing, 3–6
committee (of networks), 21
complex models, see model, complexity of
computational expense
    of Bayesian learning using Gaussian approximation, 88, 152
    of Bayesian learning using hybrid Monte Carlo, 87–88, 152
    of cross-validation, 13
    of rejection sampling from the prior, 19
computational performance
    on Boston housing data, 134
    on forensic glass data, 138
    on LED display problem, 119
    on robot arm problem, 87–88, 113, 123, 125
conjugate prior, 67
covariance function, 37, 146
cross validation, 13, 119, 127, 129

DELVE project, 152
derivatives
    erroneous computation of, 73
    of log likelihood, 162–164
    of log prior density, 162
    of potential energy, 58, 70, 93
detailed balance, 24, 27
dissipation of energy, 78
dogs, weights of, 7

domain of attraction, 43, 159

early stopping, 112
energy function, 27, 57–58, 68
    approximations to, 92
entropy-based priors, 15
equilibrium distribution, 24
    confirming convergence to, 81, 87, 106, 143
    getting close to, 76
error on training cases, 12
estimator, 3
    bias and variance of, 8, 9
    MAP, 6
    maximum likelihood, 4
    penalized likelihood, 4
evaluation of learning methods, 99–100, 126–127, 152
evidence approach, 20, 86, 108, 114
    criticism of, 20
extrapolation, 52

forensic glass data, 136
    computational performance on, 138
    neural network models for, 137–138
    predictive performance on, 139–143
    Ripley's results on, 139
fractional Brownian functions, 39–40
    hyperparameter controlling, 52
    with $\eta < 1$, 49
free energy (of window), 95
frequentist statistics, 3

Gamma distribution, 39, 67, 101, 156
Gaussian approximation, see posterior distribution, Gaussian approximation to
Gaussian distribution, 4, 21, 27, 43, 159
    example of sampling from, 62

Gaussian process, 31–42, 146
  Brownian, 38
  convergence to, 33, 36
  covariance function for, 37, 146
  direct implementation of, 43, 146
  fractional Brownian, 39–40, 146
  smooth, 38
Gibbs sampling, 25–26
  ergodicity of, 26
  for neural network model, 26
  invariance for, 25
  use in Bayesian inference, 26
  use in hybrid Monte Carlo, 60
  use in stochastic dynamics, 59

Hamiltonian dynamics, 58–59
  invariance of canonical distribution under, 59
  simulation of, 59, 61
Hamiltonian function, 57
handwriting recognition, 8
heatbath method, see Gibbs sampling
heteroscedasticity, 66, 128
hidden features, 11, 34, 43, 45, 50, 146
hidden layers
  infinite number, 50–51, 147
  more than one, 48–51, 147
hidden unit, 10, 153
  step function, 31, 35, 37, 46, 48
  tanh, 30, 36, 38, 46
hierarchical models, 6, 14, 51–53, 147–150
  as alternative to comparing several models, 127, 148
  determining input relevance, 16
  finding additive structure, 148
  other uses of, 52, 150
hybrid Monte Carlo, 56, 60–63, 150
  compared with other methods, 62–63, 88–91
  demonstration on robot arm data, 76–84

ergodicity of, 61
for bivariate Gaussian, 62–63
for Gaussian process model, 151
for neural network model, 64–66, 68–73
invariance for, 61, 98
other variants of, 151
with partial gradients, 91–95
with persistence, 71, 97–98
with windows, 95–96, 118, 138
with windows and partial gradients together, 96–97, 106, 123, 131, 134
hyperbolic tangent (tanh), 11, 75
hyperparameters, 6, 20, 156
  common, 67, 102, 115, 156
  controlling noise level, 12, 68
  controlling prior variance, 14, 66, 101
  Gibbs sampling for, 67, 83, 159–161
  in additive models, 149
  in ARD models, 16, 102
  integration over, 20
  maximization with respect to, 20
  other ways of handling, 65

infinite networks, 15, 17, 30, 103, 145, 147
initial distribution, 23
initial phase, 76–79, 102, 106, 118
input unit, 10, 153
invariant distribution, 24
irrelevant inputs, 15

kriging, 146
Kullback-Leibler divergence, 22

Langevin Monte Carlo, 61–63
  compared with hybrid Monte Carlo, 63, 88
large networks, 102–113
lattice field theory, 56
leapfrog method, 59–60

180 Index

for simple system, 71
stability of, 62, 71, 79
with individual stepsizes, 70
learning
    about parameters, 4
    Bayesian, 3, 4, 13, 17, 18
    for neural networks, 12, 13
    frequentist, 3
    in daily life, 1
    theories of, 1
LED display problem, 116
    Breiman's results on, 117
    computational performance on, 119
    neural network models for, 117–118
    predictive performance on, 119–120
likelihood function, 4, 5, 13, 19
local minima, 13, 21
logistic regression models, 155
loss function, 5

Markov chain, 23
    construction of, 25
    describing prior of infinite-layer network, 51
    ergodic, 24
    reversible, 24
Markov chain Monte Carlo, 22–28
    for neural network model, 55–98
    reviews of, 23
masses, 58, 70
    relation to stepsizes, 70
maximum *a posteriori* probability (MAP) estimate, 6, 111
maximum likelihood, 4, 8, 12
    for network applied to robot arm problem, 111–112
maximum penalized likelihood, 4, 6, 13, 111
median (guessing), 104, 106
method of sieves, 9
Metropolis algorithm, 26–28
    compared with hybrid Monte Carlo, 63, 88
    ergodicity of, 27
    for neural network model, 28
    invariance for, 27
    use in hybrid Monte Carlo, 61
Minimum Description Length, 22
mixture models, 9
ML-II, 20
model (probabilistic), 3
    based on multilayer perceptron, 12, 149
    complexity of, 2, 7–9, 21, 51, 103, 145
    hierarchical, 6
    nonparametric, 10, 30
    parameters of, 3
    posterior probability of, 148
model parameter, *see* parameters
momentum variable, 57, 69
Monte Carlo estimate, 17, 23
    based on dependent sample, 24
    for mean of predictive distribution, 64, 85
    for median of predictive distribution, 106
    for predictive distribution, 20
    variance of, 24, 85
multi-fractals, 146
multi-leap, 92
multilayer perceptron, 10, 153
    approximations using, 11, 30
    models defined using, 12, 155
    posterior distribution for, 19, 64
    prior distributions for, 14–15, 29, 53
multiple inputs, 40, 42
multiple outputs, 33, 34, 45

neural networks, 10
    applications of, 2, 12
    as models of the brain, 2
    large vs. small, 46–48, 103–104
    multilayer perceptron, 10, 153
    noise level, 12, 66, 155

for robot arm problem, 76
   prior distribution for, 68, 158
non-Gaussian stable process, 43–48
   convergence to, 44
nonparametric models, 10
normalization of inputs, 115–116, 128, 137

Occam's Razor, 2, 7, 9
offset (for a unit), 155
   prior distribution for, 158
on-line learning, 91
output unit, 11, 153
overfitting, 8, 13, 30, 103, 104, 108, 112–113, 135

parameters (of a model), 3, 6
   for a multilayer perceptron, 11, 64
partial gradients, *see* hybrid Monte Carlo, with partial gradients
performance, *see* computational performance; predictive performance
persistence, *see* hybrid Monte Carlo, with persistence
phase space, 57
   preservation of volume, 59, 60
philosophy of induction, 2, 7, 9
Poisson process, 45, 147
polynomial models, 9
position variable, 57, 68
posterior distribution, 5, 17, 19
   expectations with respect to, 23
   for neural network model, 13, 19, 64
   Gaussian approximation to, 19–22, 55–56, 151
   modes of, 19–22, 150, 151
precision values, 67, 101, 156
prediction
   Bayesian, 5, 6
   frequentist, 4, 6
   uncertainty of, 6, 9, 18, 108
   using weighted average, 21

predictive distribution, 5, 6, 14
   for Gaussian process model, 33, 146
   for neural network model, 13, 19, 20, 33, 64, 84, 108
   found using Markov chain Monte Carlo, 64, 84–87
   median of, 104, 106
   visualizing, 84
predictive performance
   on Boston housing data, 131, 132, 134–136
   on forensic glass data, 139–143
   on LED display problem, 119–120
   on robot arm problem, 85–87, 104, 107, 125
prior distribution, 4, 5
   Cauchy, 105
   choice of, 15, 51
   combined Gaussian and non-Gaussian, 49
   for a multilayer perceptron, 14–15, 29–53, 156–158
   Gaussian, 14, 16, 17, 31, 104
   improper, 7, 75, 105
   limit for infinite network, 32–34, 36, 43–45
   meaning of, 15, 29
   non-Gaussian, 16, 43, 44, 104, 147
   random generation from, 17, 18, 30, 36, 45, 147
   scaling with number of units, 32, 44, 75, 159
   vague, 7, 17, 67, 105, 114, 137, 143, 150
probabilistic model, *see* model
proposal distribution, 26–27, 61, 90

quantum chromodynamics, 56

random walks (problem of), 27, 28, 62–63, 79, 89, 97, 150
redundancy in training set, 92

regression models, 12, 14, 17, 31, 155
regularization, 4, 52
rejection rate, 72, 73, 81, 90, 91, 96–98
rejection sampling, 19
  for high-level hyperparameters, 160
  for posterior of network model, 19, 74, 166–167
robot arm problem, 75
  computational performance on, 87–88, 113, 123, 125
  demonstration of implementation on, 76–84
  large networks applied to, 104–113
  MacKay's results on, 86, 88, 108
  maximum likelihood applied to, 111–112
  neural network models for, 75–76, 104–106, 122
  predictive performance on, 85–87, 104, 107, 125
  tests of ARD on, 122–125

sampling phase, 76, 81–84, 102, 106, 118
second derivatives
  of log likelihood, 164
  of log posterior density, 19, 164
  of log prior density, 166
  of potential energy, 72
sigma values, 156
simulated annealing, 26, 65, 143
smart Monte Carlo, 62
smooth functions, 14, 15, 36, 37
smoothing splines, 146
softmax model, 12, 14, 117, 138, 155
software implementing Bayesian neural network learning
  demonstration of, 74–88
  design decisions for, 65–66
  details regarding, 153–167
  how to obtain, 169
  verifying correctness of, 73–74
squared error loss, 5, 14, 17, 84, 104, 106
stable distributions, 43, 159
stationary distribution, *see* invariant distribution
statistical physics, 22, 26
step function, 31, 35
stepsize, 60, 61
  for Langevin Monte Carlo, 63
  heuristic choice of, 72–73, 164–166
  relation to masses, 70
  selection of, 62, 66, 71–73
stepsize adjustment factor, 72, 106
  choice of, 77, 79, 133
stochastic dynamics, 58–60
  compared with hybrid Monte Carlo, 90–91
  ergodicity of, 59
  for neural network model, 65
  systematic error in, 60
structural risk minimization, 9
super-transitions, 77, 102

t-distribution, 44, 52, 101, 128, 156, 157
targets, 12, 64, 155
  binary, 156
  discrete, 12, 156
  real-valued, 12, 155
temperature, 57
tempering, 143, 151
test case, 13, 64
tests of performance, 99–100, 126–127, 152
time (fictitious), 58
timing figures, 74, 102
training cases, 12, 64
trajectory, 59
  computed using partial gradients, 92–93
  error in H along, 62, 79

optimal length of, 59, 62, 79–81, 106, 133
variation of quantities along, 79, 106
transition probabilities, 23
tuning (of implementation), 66

underfitting, 13, 103, 108

vague prior, *see* prior distribution, vague
validation set, 13–14, 112, 113

weight (on a connection), 10, 155
  prior distribution for, 157
weight decay, 13, 14, 113
width values, 157
windows, *see* hybrid Monte Carlo, with windows

Lecture Notes in Statistics

For information about Volumes 1 to 67
please contact Springer-Verlag

Vol. 68: M. Taniguchi, Higher Order Asymptotic Theory for Time Series Analysis. viii, 160 pages, 1991.

Vol. 69: N.J.D. Nagelkerke, Maximum Likelihood Estimation of Functional Relationships. V, 110 pages, 1992.

Vol. 70: K. Iida, Studies on the Optimal Search Plan. viii, 130 pages, 1992.

Vol. 71: E.M.R.A. Engel, A Road to Randomness in Physical Systems. ix, 155 pages, 1992.

Vol. 72: J.K. Lindsey, The Analysis of Stochastic Processes using GLIM. vi, 294 pages, 1992.

Vol. 73: B.C. Arnold, E. Castillo, J.-M. Sarabia, Conditionally Specified Distributions. xiii, 151 pages, 1992.

Vol. 74: P. Barone, A. Frigessi, M. Piccioni, Stochastic Models, Statistical Methods, and Algorithms in Image Analysis. vi, 258 pages, 1992.

Vol. 75: P.K. Goel, N.S. Iyengar (Eds.), Bayesian Analysis in Statistics and Econometrics. xi, 410 pages, 1992.

Vol. 76: L. Bondesson, Generalized Gamma Convolutions and Related Classes of Distributions and Densities. viii, 173 pages, 1992.

Vol. 77: E. Mammen, When Does Bootstrap Work? Asymptotic Results and Simulations. vi, 196 pages, 1992.

Vol. 78: L. Fahrmeir, B. Francis, R. Gilchrist, G. Tutz (Eds.), Advances in GLIM and Statistical Modelling: Proceedings of the GLIM92 Conference and the 7th International Workshop on Statistical Modelling, Munich, 13-17 July 1992. ix, 225 pages, 1992.

Vol. 79: N. Schmitz, Optimal Sequentially Planned Decision Procedures. xii, 209 pages, 1992.

Vol. 80: M. Fligner, J. Verducci (Eds.), Probability Models and Statistical Analyses for Ranking Data. xxii, 306 pages, 1992.

Vol. 81: P. Spirtes, C. Glymour, R. Scheines, Causation, Prediction, and Search. xxiii, 526 pages, 1993.

Vol. 82: A. Korostelev and A. Tsybakov, Minimax Theory of Image Reconstruction. xii, 268 pages, 1993.

Vol. 83: C. Gatsonis, J. Hodges, R. Kass, N. Singpurwalla (Editors), Case Studies in Bayesian Statistics. xii, 437 pages, 1993.

Vol. 84: S. Yamada, Pivotal Measures in Statistical Experiments and Sufficiency. vii, 129 pages, 1994.

Vol. 85: P. Doukhan, Mixing: Properties and Examples. xi, 142 pages, 1994.

Vol. 86: W. Vach, Logistic Regression with Missing Values in the Covariates. xi, 139 pages, 1994.

Vol. 87: J. Müller, Lectures on Random Voronoi Tessellations.vii, 134 pages, 1994.

Vol. 88: J. E. Kolassa, Series Approximation Methods in Statistics. Second Edition, ix, 183 pages, 1997.

Vol. 89: P. Cheeseman, R.W. Oldford (Editors), Selecting Models From Data: AI and Statistics IV. xii, 487 pages, 1994.

Vol. 90: A. Csenki, Dependability for Systems with a Partitioned State Space: Markov and Semi-Markov Theory and Computational Implementation. x, 241 pages, 1994.

Vol. 91: J.D. Malley, Statistical Applications of Jordan Algebras. viii, 101 pages, 1994.

Vol. 92: M. Eerola, Probabilistic Causality in Longitudinal Studies. vii, 133 pages, 1994.

Vol. 93: Bernard Van Cutsem (Editor), Classification and Dissimilarity Analysis. xiv, 238 pages, 1994.

Vol. 94: Jane F. Gentleman and G.A. Whitmore (Editors), Case Studies in Data Analysis. viii, 262 pages, 1994.

Vol. 95: Shelemyahu Zacks, Stochastic Visibility in Random Fields. x, 175 pages, 1994.

Vol. 96: Ibrahim Rahimov, Random Sums and Branching Stochastic Processes. viii, 195 pages, 1995.

Vol. 97: R. Szekli, Stochastic Ordering and Dependence in Applied Probability. viii, 194 pages, 1995.

Vol. 98: Philippe Barbe and Patrice Bertail, The Weighted Bootstrap. viii, 230 pages, 1995.

Vol. 99: C.C. Heyde (Editor), Branching Processes: Proceedings of the First World Congress. viii, 185 pages, 1995.

Vol. 100: Wlodzimierz Bryc, The Normal Distribution: Characterizations with Applications. viii, 139 pages, 1995.

Vol. 101: H.H. Andersen, M.Højbjerre, D. Sørensen, P.S.Eriksen, Linear and Graphical Models: for the Multivariate Complex Normal Distribution. x, 184 pages, 1995.

Vol. 102: A.M. Mathai, Serge B. Provost, Takesi Hayakawa, Bilinear Forms and Zonal Polynomials. x, 378 pages, 1995.

Vol. 103: Anestis Antoniadis and Georges Oppenheim (Editors), Wavelets and Statistics. vi, 411 pages, 1995.

Vol. 104: Gilg U.H. Seeber, Brian J. Francis, Reinhold Hatzinger, Gabriele Steckel-Berger (Editors), Statistical Modelling: 10th International Workshop, Innsbruck, July 10-14th 1995. x, 327 pages, 1995.

Vol. 105: Constantine Gatsonis, James S. Hodges, Robert E. Kass, Nozer D. Singpurwalla(Editors), Case Studies in Bayesian Statistics, Volume II. x, 354 pages, 1995.

Vol. 106: Harald Niederreiter, Peter Jau-Shyong Shiue (Editors), Monte Carlo and Quasi-Monte Carlo Methods in Scientific Computing. xiv, 372 pages, 1995.

Vol. 107: Masafumi Akahira, Kei Takeuchi, Non-Regular Statistical Estimation. vii, 183 pages, 1995.

Vol. 108: Wesley L. Schaible (Editor), Indirect Estimators in U.S. Federal Programs. viii, 195 pages, 1995.

Vol. 109: Helmut Rieder (Editor), Robust Statistics, Data Analysis, and Computer Intensive Methods. xiv, 427 pages, 1996.

Vol. 110: D. Bosq, Nonparametric Statistics for Stochastic Processes. xii, 169 pages, 1996.

Vol. 111: Leon Willenborg, Ton de Waal, Statistical Disclosure Control in Practice. xiv, 152 pages, 1996.

Vol. 112: Doug Fischer, Hans-J. Lenz (Editors), Learning from Data. xii, 450 pages, 1996.

Vol. 113: Rainer Schwabe, Optimum Designs for Multi-Factor Models. viii, 124 pages, 1996.

Vol. 114: C.C. Heyde, Yu. V. Prohorov, R. Pyke, and S. T. Rachev (Editors), Athens Conference on Applied Probability and Time Series Analysis Volume I: Applied Probability In Honor of J.M. Gani. viii, 424 pages, 1996.

Vol. 115: P.M. Robinson, M. Rosenblatt (Editors), Athens Conference on Applied Probability and Time Series Analysis Volume II: Time Series Analysis In Memory of E.J. Hannan. viii, 448 pages, 1996.

Vol. 116: Genshiro Kitagawa and Will Gersch, Smoothness Priors Analysis of Time Series. x, 261 pages, 1996.

Vol. 117: Paul Glasserman, Karl Sigman, David D. Yao (Editors), Stochastic Networks. xii, 298 pages, 1996.

Vol. 118: Radford M. Neal, Bayesian Learning for Neural Networks. xv, 183 pages, 1996.

Vol. 119: Masanao Aoki, Arthur M. Havenner, Applications of Computer Aided Time Series Modeling. ix, 329 pages, 1997.

Vol. 120: Maia Berkane, Latent Variable Modeling and Applications to Causality. vi, 288 pages, 1997.

Vol. 121: Constantine Gatsonis, James S. Hodges, Robert E. Kass, Robert McCulloch, Peter Rossi, Nozer D. Singpurwalla (Editors), Case Studies in Bayesian Statistics, Volume III. xvi, 487 pages, 1997.

Vol. 122: Timothy G. Gregoire, David R. Brillinger, Peter J. Diggle, Estelle Russek-Cohen, William G. Warren, Russell D. Wolfinger (Editors), Modeling Longitudinal and Spatially Correlated Data. x, 402 pages, 1997.

Vol. 123: D. Y. Lin and T. R. Fleming (Editors), Proceedings of the First Seattle Symposium in Biostatistics: Survival Analysis. xiii, 308 pages, 1997.

Vol. 124: Christine H. Müller, Robust Planning and Analysis of Experiments. x, 234 pages, 1997.

Vol. 125: Valerii V. Fedorov and Peter Hackl, Model-oriented Design of Experiments. viii, 117 pages, 1997.

Vol. 126: Geert Verbeke and Geert Molenberghs, Linear Mixed Models in Practice: A SAS-Oriented Approach. xiii, 306 pages, 1997.

Vol. 127: Harald Niederreiter, Peter Hellekalek, Gerhard Larcher, and Peter Zinterhof (Editors), Monte Carlo and Quasi-Monte Carlo Methods 1996, xii, 448 pages, 1997.

Vol. 128: L. Accardi and C.C. Heyde (Editors), Probability Towards 2000, x, 356 pages, 1998.

Vol. 129: Wolfgang Härdle, Gerard Kerkyacharian, Dominique Picard, and Alexander Tsybakov, Wavelets, Approximation, and Statistical Applications, xvi, 265 pages, 1998.

Vol. 130: Bo-Cheng Wei, Exponential Family Nonlinear Models, ix, 240 pages, 1998.

Vol. 131: Joel L. Horowitz, Semiparametric Methods in Econometrics, ix, 204 pages, 1998.

Vol. 132: Douglas Nychka, Walter W. Piegorsch, and Lawrence H. Cox (Editors), Case Studies in Environmental Statistics, viii, 200 pages, 1998.

Vol. 133: Dipak Dey, Peter Müller, and Debajyoti Sinha (Editors), Practical Nonparametric and Semiparametric Bayesian Statistics, xv, 408 pages, 1998.

Vol. 134: Yu. A. Kutoyants, Statistical Inference For Spatial Poisson Processes, vii, 284 pages, 1998.

Vol. 135: Christian P. Robert, Discretization and MCMC Convergence Assessment, x, 192 pages, 1998.

Vol. 136: Gregory C. Reinsel, Raja P. Velu, Multivariate Reduced-Rank Regression, xiii, 272 pages, 1998.

Vol. 137: V. Seshadri, The Inverse Gaussian Distribution: Statistical Theory and Applications, xi, 360 pages, 1998.

Vol. 138: Peter Hellekalek, Gerhard Larcher (Editors), Random and Quasi-Random Point Sets, xi, 352 pages, 1998.

Vol. 139: Roger B. Nelsen, An Introduction to Copulas, xi, 232 pages, 1999.

Vol. 140: Constantine Gatsonis, Robert E. Kass, Bradley Carlin, Alicia Carriquiry, Andrew Gelman, Isabella Verdinelli, Mike West (Editors), Case Studies in Bayesian Statistics, Volume IV, xvi, 456 pages, 1999.

Vol. 141: Peter Müller, Brani Vidakovic (Editors), Bayesian Inference in Wavelet-Based Models, xi, 394 pages, 1999.